THE Prenatal Prescription

THE Prenatal Prescription

PETER NATHANIELSZ

WITH

CHRISTOPHER VAUGHAN

A LIVING PLANET BOOK

HarperCollins*Publishers*

THE PRENATAL PRESCRIPTION. Copyright © 2001 by Peter Nathanielsz and Christopher Vaughan. All rights reserved. Printed in the United States of America. No part of this book may be used or reproduced in any manner whatsoever without written permission except in the case of brief quotations embodied in critical articles and reviews. For information, address HarperCollins Publishers Inc., 10 East 53rd Street, New York, NY 10022.

HarperCollins books may be purchased for educational, business, or sales promotional use. For information, please write: Special Markets Department, HarperCollins Publishers Inc., 10 East 53rd Street, New York, NY 10022.

FIRST EDITION

Designed by Jackie McKee

Printed on acid-free paper

Library of Congress Cataloging-in-Publication Data is available upon request.

ISBN 0-06-019763-3

01 02 03 04 05 ❖ RRD 10 9 8 7 6 5 4 3 2 1

For Tomorrow's Children

Contents

Acknowledgments

THIS BOOK RELATES THE WORK OF MANY OUTSTANDING researchers who have revealed the mysteries of life in the womb. I have been fortunate to know most of them personally. In order to keep the flow of the text, I have resisted the urge to name them at each occasion their work is discussed. I am particularly indebted for the many hours of discussion I have had over the years with dedicated researchers all over the world who are trying to understand the underlying principles and mechanisms that make programming a feature of the biology of all mammalian species. Some of these are clinicians, and some are basic scientists: David Barker, Otto Blekker, John Challis, James Clapp, Gautam Chaudhuri, Jorge Figueroa, Bert Garza, Peter Gluckman, Keith Godfrey, Robert Goldenberg, Dino Giussani, Nicholas Hales, Mark Hanson, Jane Harding, Michael Heymann, Barbera Honnebier, Thomas Kirschbaum, Janna Koppe, Mont Liggins, Charles Lockwood, Larry Longo, Stephen Lye, Thomas McDonald, Majid Mirmiran, Murray Mitchell, Leslie Myatt, Mark Nijland, William Oh, David Olson, Lucilla Poston, Kathie Rasmussen, Jeffrey Robinson,

Roberto Romero, Maria Seron-Ferre, "Buddy" Stark, Dick Swaab, Andre Van Assche, Charles Wood, Wen Wu, and many, many others. All are dedicated to a better understanding of the fetal origins of adult disease in order to provide a better life for future generations. I am grateful for their enthusiasm, learning, and energy. Researchers in this field know that we could not study these important events without the tireless help of those who direct the various programs of the National Institutes of Health in the United States and equivalent bodies such as the Wellcome Trust in the United Kingdom. This work would not be possible without the financial and intellectual support that comes from society through these institutions. I would like to add my own very special thanks to those smaller charities and foundations who help young researchers to get started. We all owe a great debt to those who work in any capacity with the more specialized granting agencies such as Tommy's Campaign, The Lalor Foundation, WellBeing, and many other such bodies. Other friends who are not among the scientific community have encouraged me in this effort to put this story before the general reader. None more so than my dear friends Lynda and Robert Lloyd, who encouraged me in my task in moments of uncertainty and when I had my doubts. To Lynda in particular I owe many thanks for ideas that have made many concepts more accessible. Last, and most importantly, this book relates a story that is central to families, and I would like to thank Diana, Julie, and David, and the members of my own first family, Aubrey, Mary, and Christine for all the support they have each and every one given me in a thousand ways.

Working with Chris Vaughan has been both a pleasure and an enriching experience. His help in getting across complex ideas in a very simple fashion has greatly helped my task. Josh Horwitz at Living Planet Books and my agent, Gail Ross, made this all happen. Finally, I want to thank Gail Winston at HarperCollins. Her enthusiam and belief in this book as not only the fascinating topic that it is but also as a subject of great importance to every family has been a source of great inspiration to me. Every author knows how important it is to have an editor who helps you fnd your voice.

Ithaca, New York
November 12, 2000

Introduction

AS PARENTS, THE GREATEST GIFT WE CAN GIVE TO OUR CHIL-dren is the gift of lifelong health. This promise is not as outlandish as it may first seem. Biomedical research over the past decades has conclusively determined that the physical, hormonal, and even emotional interaction between a mother and her child in the womb has a concrete effect on that childs physical and mental health for decades to come. This discovery—*the single most important story to come from biomedical research since the determination of the structure of the gene*—is the subject of this book. Sometimes called "fetal programming," or "the fetal origins of adult disease," this research suggests that the long-term effects of the environment in the womb on the health we enjoy throughout life are even more important than our genes. They are certainly much more under our control.

The fetal programming story is one of optimism and hope, because there are things that we can do to ensure that tomorrow's children—including your own precious baby—have the best possible home before birth and therefore have the best chance of a healthy life. And for those of us who have already passed through the first important chapters of life, there is still a benefit in understanding this exciting new knowledge. Understanding the implications of fetal programming enable all of us to live a lifestyle that is more in tune with who we are, biologically.

For over thirty years I have been striving to understand the beautiful and intriguing mysteries of life in the womb. I have spent much of my time trying to delve into the intimate hormonal and chemical conversations that pass between a mother and her child during the nine months of gestation. From my early days at Cambridge University to my current position as director of the Laboratory for Pregnancy and Newborn Research at Cornell University, I have been promoting child and maternal well-being by investigating how we can best nurture our children's health. At home, my wife and I have learned that much of what we do in rearing our children is done by trial and error. You may read the books and learn all you can about getting them to eat a healthy diet and behave properly, but when you are actually parenting a child, you often find yourself making it up as you go along. But at the same time, in my laboratory and through interaction with other researchers at scientific conferences around the world, I was learning that there is another type of parenting that we can start much earlier and that has long-lasting benefits. I was learning how we can parent children prenatally so that they develop in the healthiest possible way. Furthermore, in contrast to the catch-as-catch-can style of parenting required when actually raising these adorable but chaotic children, prenatal parenting can be done with a fairly simple and doable program.

What I want to do is show you how and why a program of prenatal parenting can work for you. First, I want us to look at how and why prenatal programming takes place, how nutrition, stress, exercise, toxins, tobacco, and other environmental influences can be factors for good or ill during your pregnancy. Second, I want to provide practical and easy tools that you can use in your own life to make your prenatal interactions with your baby memorable and as beneficial as possible.

How do we know that life before birth strongly influences our health for decades? The evidence is impressive and overwhelming, and comes from three very different sources, each of which reinforces the others: human epidemiological data, research with patients, and animal research. The first source of evidence grows out of massive human epidemiologic databases of medical records of babies born in the last century. Researchers have followed these people throughout their lives to see what sort of health they enjoyed, and how their lifetime health patterns correlated with their weight at birth and other indications of the quality of life they enjoyed in the womb. In doing this, these intrepid and hardworking researchers have amassed impressive, and compelling, reservoirs of modern medical data that reveal the nature of prenatal programming.

Next is the evidence that comes from animal studies conducted in many laboratories throughout the world. The crucial strength of the animal studies is that they are carefully controlled and can be repeated and confirmed independently by researchers with different biases. While some people might claim that "research on rats is not the same as what happens in people," animals and humans use very similar underlying common mechanisms in the care and feeding of their young. In the case of the placenta, for instance, every placenta—whether human, hyena, or hippo—must transfer oxygen one way and carbon dioxide the other. We have a saying in veterinary research that there is "one medicine," one system for understanding health and disease in all animals, including the human animal. Regardless of whether the information comes from a human, a horse, or a rat, the information always tells us something about how our common biological mechanisms work.

Finally, the concept of programming of lifetime health is further supported by clinical research on patients with a wide variety of diseases. By tracing back the origins of many diseases, physicians are more and more often finding that the origins of such diseases lie in fetal life. So the overwhelming evidence that the quality of life in the womb is critical to lifelong health comes from this "triangle of evidence" that cries out to us to pay attention.

Fetal programming, one of those few scientific ideas that comes along and completely alters conventional wisdom about the best ways to rear healthy, happy children, is truly a paradigm shift. Like many revolutionary ideas, it started out as a seeming fantasy because it was so at odds with what was scientifically known and accepted. After years of research and questioning, the idea has slowly become accepted among scientists.

The scientific interest in the issue of prenatal programming is now so intense that in the last year I have attended three meetings of the National Institutes of Health that have addressed the issue. The interest is so wide-ranging that one of the meetings was convened by no less than five of the important Institutes and Centers at the NIH. The National Institute of Child Health and Human Development, directed by Dr. Duane Alexander, has perhaps the most direct interest in fetal development, so it was not surprising that it should convene a workshop on the topic. The other cosponsors also had a clear interest in fetal programming. The National Heart, Lung, and Blood Institute has noted the important observation that low birth weight is associated with heart disease and stroke in later life. The National Institute of Diabetes and Digestive and Kidney Diseases was interested not only to learn more about the studies indicating a relationship between poor maternal nutrition during pregnancy and the likelihood that children will get diabetes in later life, but also how the increased likelihood of getting diabetes can be passed down intergenerationally from mother to daughter and granddaughter. The Office of Research on Women's Health, directed by Dr. Vivian Pinn, has taken a deep interest in the concept of fetal programming as well. From the biological point of view, the story of fetal programming shows clearly that for the health of our society, women's health is more important than mens. As we will see, many aspects of good health, not just a tendency to diabetes, are passed from mother to daughter. For men, the quality of life in the womb is important for their own health, but they do not pass it on down through following generations.

Perhaps the most remarkable expression of scientific and government interest about fetal development was the involvement of the National Institute on Aging in this meeting. Gerontologists traditionally don't spend much time studying prenatal development. It is now very clear, however, that how we age is determined in profound ways by how well the placenta functions in the womb. After a session on the placenta, I stood in the coffee line talking about this fascinating new information with an extremely interested official from the National Institute on Aging and realized what a true paradigm shift had taken place.

Despite the now-fervid interest of the scientific community in prenatal programming, most parents and parents-to-be still know little or nothing about the issue. It's high time that changed. Parents need to get the message that they can instill better health in their children through positive prenatal programming. I am gratified that when I speak to audiences around the world, they understand immediately the importance of the new science of life in the womb. Rather than be alarmed by the news, people generally find it encouraging to hear that given the right opportunities and approach to pregnancy, families can both improve lifetime health and lower the risk of serious disease throughout their child's life. Parents can even benefit future generations, since healthy daughters have healthier children when they themselves become pregnant.

In addition to speaking publicly, I have tried to work directly with mothers and fathers to show them how they can integrate prenatal programming into their pregnancy. I will admit that, as with many other kinds of advice about health, incorporating the ideals of prenatal programming into ones daily life is not always easy.

One woman with whom I was working was a very successful advertising executive for a major national magazine based in Manhattan. She was the kind of person who utterly threw herself into her high-pressure job and lived with the stress. She had been devastated when her first pregnancy had gone badly and ended with a premature delivery associated with some major complications. When I first met her, she was trying to get pregnant again and wanted advice on how to avoid repeating the terrible course of the first pregnancy. I asked about her work and her life, trying to get a sense of how she ate, exercised, slept, and how her work made her feel (always under pressure, always stressed, she told me). In the end I told her that her next baby would benefit if she cut back on work a little during the pregnancy, worked on techniques to manage the stress that her work did create, and increased blood flow to her womb by lying down regularly in her office. Luckily she was receptive to my advice. When she became pregnant, and after some worry about how her colleagues would take her actions, she threw herself into creating a personal program for prenatal health with as much gusto as she took on any project. Her husband also was able to understand the nature of prenatal programming and was supportive of changes she made to take it easier at home. Actually, the person who most resisted easing off at home and work was herself. In the end she realized that the changes I was suggesting were temporary, but the results would last all her child's life. As it turned out, her baby was born healthy, at full term. Although we wont know for many years how this particular child fares, the mass of scientific research assures us that she did make a difference in his life.

In the end, however, I know that my efforts at public speaking, and my working with couples and individual women, reaches only a small percentage of the people who could benefit from the knowledge of prenatal programming. That is why I am writing this book. In the following pages, I have tried to give you the information you need to tailor your own pregnancy program. After reading *The Prenatal Prescription*, you and your spouse will understand how to evaluate your own health status, diet, exercise, and stress levels. You will understand how preparation of the womb begins long before pregnancy, and how you can best prepare yourself. In creating a program for pregnancy, I hope you will be guided by four principles: knowledge, simplicity, pragmatism, and optimism. I have highlighted these qualities because

Knowledge is the beginning of all power. Before we can do anything, we have to know what to do. As a university professor, I have observed that people who understand something are much more likely to do it well and enjoy doing it. In addition, knowledge of the hows and whys of fetal programming helps you tailor a prenatal parenting program to your own needs—because each pregnancy is as different as the mother and baby involved.

Simplicity is important because our lives are already overwhelmingly complex, and pregnancy introduces a whole new level of logistical complexity. While the science of life in the womb is not easy to implement, its most important lessons for expectant mothers can be distilled into simple steps that have profound lifelong benefits.

Pragmatism is the necessary complement of simplicity. No one can put their lives on hold for nine months—nor should they. Women have jobs, family, and other important commitments that can't be totally subordinated to their pregnancy. It is not possible for every pregnant woman to eat perfectly at every meal. Nor can anyone shield themselves from all sources of stress—pregnancy itself creates stressful challenges—or from every environmental toxin. In the real world, life goes on even during pregnancy. The human species is a very resilient species—we are not at the top of the biological tree for nothing. We can still do quite well even when we have to adapt to practical realities. It is important to remember that pregnancy is normally normal. Challenge and stress will always be with your children as they grow. Sir Isaac Newton, arguably the greatest physicist and mathematician of all time, is said to have been so small at birth that you could have squeezed him into a pint pot. Well, he did quite well, didn't he?

Optimism, the belief that things are going to turn out well, and that you can accomplish your goals, is essential to making positive changes in your life. If knowledge gives you the foundation for change, optimism gives you the strength to build on that foundation. To me, the new science of life in the womb is tremendously encouraging and exciting. The science itself is promising, but what really gets me going is the fact that pregnant women can take the lead in putting its practical consequences into action. However, all sectors of society—teachers, politicians, economists, physicians, indeed everyone—must realize that there is much that can, and should, be done to improve the quality of life before birth for their own and all of tomorrow's children.

Above all, remember that you are trying to improve your baby's environment in the womb, not make it perfect. It is counterproductive and needlessly guilt producing to aspire to an unachievable standard of perfection. Each of us is only too aware of our limitations, and of the fact that our actions often fall short of the ideal we hold up for ourselves. We don't need to produce more guilt and worry during a time that should be one of the most profound and satisfying you ever experience.

What you should strive to do is achieve a balance of prudence and pragmatism. I hope that you and your partner, your extended family, and your circle of friends, will use the information on the importance of fetal development to improve the life and livelihood of your growing child. I hope that you and your spouse will read this book together, since now more than ever couples are partners who need to learn together and work together as they begin the marvelous task of parenting children from the very earliest time. As I always say, "If you fail to prepare, you prepare to fail."

Whether you are already pregnant or preparing for pregnancy, you will know how to make the small but crucial changes that will allow you to not only give birth to healthy babies, but to give your children a gift that stays with them for the rest of their lives. In all things it is important to remember that it is never too early, or too late, to start. It is also important to remember that there is no "perfect baby." You cannot change your baby's genes, but you can make sure that your baby gets the best start in life. What could be a deeper expression of love? What could be more empowering and exciting?

1

Prenatal Programming:
The New Science of Life in the Womb

IMAGINE A WARM SOUTHERN CALIFORNIA EVENING IN AUGUST 1921. Alice Miller is preparing dinner when she begins to feel the bands of muscle in her womb tighten in earnest. For the last few nights, these muscles have been more active than usual, but never contracting this strongly or for this long. Nurses sponge her brow through a long night of labor, and as the sun comes up on a bright summer morning, her first child is born. She names him James. James is a chubby, healthy boy right from the start, and from the moment Alice first swaddles him in white cotton and gently cradles his little pink body in her arms, she senses he will be a lucky child. As he grows he proves to be both lucky and happy, able to weather well the normal knocks of childhood. His parent's luck doesn't hold, however. A year after his birth his father, Michael, an engineer at an electric plant in the San Fernando Valley, is hit by a truck, disabled, and has to leave the well-paid job he has held for close to a decade. The Miller family is forced to move back to their native Pittsburgh, where they can be nearer the support of family and friends. James's mother keeps the family fed by working at a large commercial laundry—much of the work is moving heavy sheets, and she is on her feet all the time. Michael becomes very depressed, and relationships among family members suffer under the financial strains. There is often not enough money for Alice to purchase the fresh fruits and vegetables and other nutritionally balanced food she bought in California. Her world becomes even more stressful five years later when she is pregnant with James's brother, William. Her income is crucial to the family, and she keeps working six days a week until her water breaks.

As James grows up and then grows old, he continues to do well, taking each new stage of life as a challenge to be overcome. Throughout childhood, James has the same diet as most everyone else in his neighborhood: a diet high in carbohydrates and fat, but one with little protein and almost no vegetables or fresh fruit. But while this diet and lifestyle produces high blood pressure and blood sugar problems in most of his neighborhood buddies as they reach middle age, James's cholesterol and blood pressure stay low. As they reach their late fifties and early sixties, many of James's friends—born in the difficult years of the depression when their mothers were not well cared for during pregnancy—begin to suffer from chronic diseases, including heart disease and diabetes. His brother, William, is diagnosed with high blood pressure at forty, then diabetes at fifty, and then dies of a stroke in his early sixties. But at seventy-nine, James has stayed fairly healthy.

What is the difference between James and his neighbors? Between James and his brother, William? How can people be exposed for most of their adult lives to the same environment and turn out so differently? This is the conundrum, the million-dollar question, that has stumped health scientists for decades. We now know that the key to this conundrum almost certainly lies largely in the disparate environments that James and William were exposed to during the crucial nine months before birth. Despite the fact that they are brothers with similar, though not identical, sets of genes, their bodies function very differently as a result of the very different conditions they experienced during prenatal life.

We have all noticed how differently people function. Some people eat almost anything, in any amounts, and never gain any weight. Their cholesterol and blood pressure stay low regardless of their diet. Other people watch what they eat, exercise regularly, and still have a tough time with their weight, blood pressure, or cholesterol. Understanding these differences has become one of the major challenges of modern health science. As one fellow scientist puts it, "We want to find out why some people are built like a Rolls-Royce and other people are built like a Yugo." For a long time, scientists and the public have attributed these differences to genes, diet, and general lifestyle. But in recent years we have garnered enough evidence to know that, while these three factors play important roles, much of the way our bodies work is molded and solidified during our life in the womb. There are critical periods during prenatal development when our cells and organs decide how they will behave for the rest of our lives. Like James Miller, we may live in one location, but our bodies are still listening to the lessons they learned in the womb, lessons taught by very different life circumstances. James lived most of his life in a working class neighborhood in Pittsburgh, but his pattern of health and sickness more closely resembled that of residents in the well-off, sunny, fruit-laden area of southern California. His body was indelibly changed long before he made any of his own choices about what kind of food to eat or how often to exercise in later life. William was less fortunate because, as a result of the situation in which his mother lived during his life in the womb, he was provided with a less supportive womb environment. This permanent molding of his basic body functions is a perfect example of fetal programming.

We get the idea for the word *programming* from a twentieth-century scientist, Konrad Lorenz. Lorenz discovered that newly hatched geese will immediately bond with the first moving animal they see after emerging from their shell. Fortunately, this first living thing is usually their mother. But Lorenz showed that goslings would just as easily bond with him and follow him everywhere if he, Lorenz, was the first being they saw. This is an example of postnatal programming, a permanent change imposed on the goslings by their environment immediately after birth.

Our bodies are programmed in analogous and far-reaching ways by our experiences before birth. The events to which are exposed as we develop our body's structure and functions during our life in the womb can improve or worsen our:

- Blood pressure
- Cardiovascular health
- Eating patterns
- Tendency to gain weight
- Emotional resilience
- Intelligence
- Susceptibility to cancer
- Resistance to infection

Consider only a few amazing findings:

- Leading researchers studying prenatal life now believe that the nutritional quality of the womb environment is often a more important predictor for risk of heart disease than either genetic predisposition or postbirth influences like diet and stress.

• Studies show that blood pressure in mothers during pregnancy correlates directly with the blood pressure of their offspring in adulthood.

• Newborns who have a disproportionate head-to-waist size, a hallmark of detrimental prenatal programming due to poor nutrition, are more likely to develop elevated levels of cholesterol as adults.

• Girls who are born with low birth weight reach sexual maturity earlier and have a smaller final stature than girls born with normal birth weight.

• Chronic maternal stress during pregnancy—both emotional and physical—can interfere with how the fetus utilizes nutrients and can affect how well or poorly a child functions psychologically throughout life.

In short, prenatal programming affects every aspect of our physical and mental health, at every stage of our lives. I like to say that how, and when, we leave this world is shaped in large part by how we entered it. For each family, this new way of thinking represents a revolution in prenatal science, a radical change that overthrows existing assumptions about how the womb environment affects a developing baby.

At first you might find these ideas scary (or perhaps appalling): they may seem like a whole new way to blame mothers for their children's problems. But apportioning blame should not be the lesson. For starters, whole families are responsible for keeping a pregnancy healthy, not just the mother. I would take this one step further. The responsibility for healthy newborn babies lies with the whole of society. Indeed, for women and their partners reading this book who are pregnant or planning on becoming pregnant, this revolutionary new understanding should bring the hopeful, encouraging news that you can you can improve your children's physical and psychological health throughout their lives by doing your very best to enrich their environment before birth.

DETECTIVE STORIES:
THE HISTORY OF PRENATAL PROGRAMMING

While prenatal programming is new to many doctors and even health scientists, a compelling body of evidence in favor of the idea has been building for years. As we have seen, the evidence has come from many separate lines of investigation, such as cellular, biochemical, and animal research, but the first systematic, dramatic, and convincing work came from the disease detectives who practice human epidemiology. And most dramatic and convincing of all has been the lifelong work of a British physician-scientist named David Barker.

Barker is a relaxed, gregarious character, and visiting him in his office on the grounds of Southampton General Hospital in England is always an enjoyable affair. Yet his relaxed manner and quiet voice camouflage a driven mind, obsessed with finding and analyzing the raw data of epidemiological research, intolerant of any scientific sloppiness. Those who spend time with him occasionally find that his eyes glaze over and gaze at some far-off place—at these times his wife notes that "he's fallen to thinking."

Barker's particular obsession of more than the last three decades has been understanding the relationship between prenatal development and lifelong health. As far back as the mid-1960s he investigated the relationship between the rate of fetal growth and intelligence. His work showed that low intelligence was generally associated with a slower rate of growth in the womb. But measuring intelligence and weeding out all the contaminating economic and social factors has always been a messy and contentious affair. Scientific revolutions have to be founded on solid ground. Epidemiology is a very important approach to understanding our health, but it has great limitations since it is impossible to find groups of humans who are exactly alike and then to change just the one thing that the researcher wishes to observe. Worse still, even if we could find such a uniform group of people, in order to see the effect of a particular factor, we would need to change the factor randomly in one half of the group while leaving the other half unchanged.

Despite these problems, Barker found solid ground several years later while looking at the connection between heart disease and prenatal growth. In 1984, Barker was editor of an atlas of England and Wales published by the Medical Research Council showing mortality rates from various diseases all over England. Barker and his colleague Clive Osmond were first struck by a somewhat heretical paradox. They noted that the areas of England and Wales that had high rates of heart disease among middle-aged and elderly men in the period 1968–1978 were the industrialized areas such as the coal mining districts. The paradox was that heart disease occurred most frequently in these underprivileged areas, not in the affluent areas around London.

The conventional wisdom at that time (and a view *still* held by many) was that heart disease was a disease of affluence. People who suffered heart attacks were thought to eat too many high-fat foods, too much red meat, and do too little exercise. If that was true, poorer areas in which people worked hard in the mines and couldn't afford rich foods should have lower rates of heart disease than the inhabitants of more affluent areas. But this simply wasn't the case. Barker was truly surprised to find that the *poorer* areas were the ones with a *higher* incidence of cardiovascular disease.

Finding the highest rates of heart disease among the under-privileged in itself broke many age-old scientific beliefs. However, an even more iconoclastic idea was to come from the same atlas: in the book lay a map from early in the first decade of the 1900s showing stillbirths and mortality in the first month of life. The areas of highest newborn mortality were very similar to the areas of highest heart disease during the period 1968–1978. Other researchers had been saying that the quality of nutrition in the months *after* birth were important in the development of heart disease. David Barker's "eureka moment" was to make the corre-lation that the quality of life even *before* birth was associated with the development of heart disease in later life. Obviously, if you were a baby who died in the first month of life in 1900, you could not die again of heart disease at the age of sixty-eight. However, David Barker surmised that if you were a disadvantaged baby in the womb and survived the first year after birth, you were more susceptible to heart disease in later life.

Barker proposed that the quality of life in the womb is a major determinant of lifetime health in general. This relationship has come to be known as the Barker hypothesis.

To prove this hypothesis, Barker had to look at the data in a new way. Obviously, it is difficult to separate the effects of poor nutrition before birth and poor nutrition immediately after birth. That is because a mother who eats poorly while pregnant is likely to eat poorly while breast-feeding, and her children will probably continue to eat poorly once breast-feeding stops. What Barker did was look at mortality rates of *newborns*, rather than infants, because any health problems in the first month of life were more likely due to prenatal rather than postnatal problems. When Barker looked at the correlation between mortality rates in the first month of life and the rates of heart disease years later, he found a close correlation. This was a very strong suggestion that difficulties in the womb, rather than in early infancy, were likely to be the more important cause of heart problems later in life. The data also showed that a healthy diet and environment after birth has an additional effect on the risk of heart disease, but it was not as important as prenatal diet and environment.

Barker and Osmond presented their findings to their scientific colleagues and were rebuffed. Most experts simply could not—or would not—believe that effects in the womb could surface decades later, and were convinced that there had to be other reasons for the remarkable correlation. Barker set out on a daunting journey that would prove the point one way or the other. What he needed was a large number of decades-old birth records. His intention was to find the birth records and then track down the babies and give these now grown-up babies a very belated postnatal checkup. With these two sets of data, Barker could test his hypothesis by comparing the subjects' lifetime health to their birth weights and other indicators of their health in the womb.

In 1985, Barker found five hundred birth records from the beginning of the twentieth century in the town of Plymouth in the southwest corner of England. Plymouth is famous as the place where Sir Francis Drake made arduous preparations to defeat the Spanish Armada. Barker's course was to prove equally arduous. He was denied access to the records he so badly wanted because they were restricted under a one-hundred-year rule of confidentiality. He was frustrated but not deterred.

The next year he discovered another complete set of birth records in six villages in Hertfordshire, a county just a few miles north of London. Again, his requests to view the records were denied because of rules about patient confidentiality. This time, however, Barker had a personal connection. He was able to tell the medical officer in the Hertfordshire records department that during World War II he and his family had been evacuated to Hertfordshire to avoid German bombing. It so happened that Barker's own sister's birth records were among those he was trying to access.

After further consideration by the Hertfordshire authorities, Barker was told that he could have the birth records if he could find a safe place to store them. That very day he telephoned the vice chancellor of Southampton University and asked if they could satisfy this requirement. With a speed that should surprise anyone who has any experience with a university bureaucracy, the vice chancellor told him that the university would guarantee the safety of the records. In fact, if the Hertfordshire authorities had any doubts, Barker was told that he could inform them that the precious records would be stored in the same place as the archives of the duke of Wellington.

Finding the first records from Hertfordshire proved to be a turning point. From 1911 onward, midwives had visited every woman who had given birth in the county. They recorded the birth weight, the way the baby was fed, any illnesses and, very importantly as we shall see, the baby's weight at one year of life. A unique strength of this record is the fact that the records were not from a selected population, such as special high- risk patients delivering in a hospital, but were a complete record of all the children born throughout the county, babies both large and small, children of affluent parents as well as of parents who were not so well off.

Armed with the knowledge that it was possible to obtain the

type of detailed birth records he would need, Barker felt the time had come to hire a researcher to help him find and sort the information. He hired a young historian recently graduated from Oxford University, and she played a key role in exploiting the Hertfordshire records. But there was another stroke of luck in her appointment. She was the first woman from her school at Preston in the north of England to study at Oxford. On one vacation from researching the database of birth records in Southampton, she went home to Preston and decided to visit her local hospital to search for further records. Her impulse was amply rewarded. There in the records room of the Sharoe Green Hospital she found the birth records from the delivery room on every baby born between 1934 and 1943. The value of these records lay in the fact that a multitude of measurements had been made on the babies at birth. The babies had been measured in such detail that Barker says he could have used the recorded measurements to make sketches of the individual babies in their cribs.

Each record at Sharoe Green contained one piece of unusual information that would open up new lines of inquiry of critical importance. The babies' birth records contained the weight of the placenta, an organ that is potentially a mine of information about the baby's stay in the womb. I have often thought that one day in the not-far-distant future the placenta will be used as the baby's prenatal diary. Then a third set of birth records showed up in the boiler house of Jessop Hospital in Sheffield, Yorkshire. These records were contained in beautifully bound ledgers that detailed every birth in the hospital dating back to 1907. These records were extremely useful in that they contained even greater details about the babies' measurements, including length and abdominal and chest circumference.

Once they had the data in hand, David Barker and his colleagues could then go to the British National Health Service and obtain records on the patients' whole health histories, including information on their visits to the doctor, their weight and blood pressure management, and on the various disorders that had afflicted them over a lifetime. They discovered some amazing correlations. One astounding and unexpected discovery was that babies with large placentas were much more likely to have high blood pressure later in life. One would be forgiven for thinking that having a large placenta would be a good sign. However, we can stand the argument on its head. It appears that when conditions are not optimal in the womb, the baby tries to grow a large placenta to compensate.

Barker and his colleagues also found that fetal growth in the womb is directly related to how well people in old age control the level of glucose in their blood. The body must carefully regulate the level of blood glucose, which is the primary source of energy for cells throughout the body. Blood glucose levels that are too high or too low can injure or kill many different types of cells in your body, leading to heart disease, visual problems, and other types of ill health. Again, an astounding correlation. Somehow these elderly citizens were getting sick not because of anything they did or the genes they had, but because their bodies had remembered a negative experience at the very beginning of life, before they were even conscious of their surroundings. It would be as if you discovered that your house, which had served you well for decades, was about to fall down from creeping rot because a factory worker had knocked a pinhole in one of the shingles that ended up on your roof.

Another interesting finding soon followed on the heels of these two major observations. For years the conventional wisdom has been that the control of blood cholesterol levels was solely a consequence of one's genetic makeup and its interaction with diet and lifestyle. There are two forms of cholesterol circulating in your blood: low-density lipoprotein cholesterol (so-called bad cholesterol) and a high-density form (good cholesterol). When Barker measured the levels of the bad cholesterol in his older subjects, he found that it correlated best not with what they ate then or at any point in their life, but with their abdominal girth at birth. People born with an abdomen that was small for their length or head size had much higher bad cholesterol even by the age of fifty than those born with a larger abdomen. This correlation seems at first glance to be bizarre, and there is an understandable tendency to believe that there must be some mistake in the research, some hidden factor that would provide a more conventional explanation for the data.

Like many scientific pioneers, Barker faced great disbelief when he tried to publish his research. When he submitted his paper on the correlation between blood pressure and placental size to a top medical journal, the editor promptly rejected it, refusing even to send it out for review. When this out-of-hand rejection happens, it is usually because the editor feels the study is so obviously unsound or absurd that publishing it will damage the journal's reputation. Barker submitted the paper to a different medical journal, and it was published, but his second paper, on prenatal effects on later blood-glucose control, was also initially rejected by the first scientific journal to which it was submitted.

Barker didn't waste much time worrying about resistance to his ideas from the scientific establishment. Even as his initial studies were published and debated, he traveled the world looking for birth records that would reveal whether or not prenatal programming affected all humans, not just Europeans. Through persistence, courage, and luck, he found extensive birth records from early in the twentieth century in India and China. Soon other researchers joined him in mining the information found in local birth records for connections to later health problems. The new studies continue to show that these connections are real, that something happens in the first days of our lives in the womb which affects us until the last days of our lives.

THE SURVIVAL OF THE FETUS

What exactly is this strange connection between body measurements at the beginning of life and the health we enjoy at end of life? How does prenatal programming work? In my mind, there are ten elements, or principles, to prenatal programming. I will explain each of these more fully in a bit, but the ten principles can be stated most simply as follows:

1. During development in the womb, there are *critical periods of vulnerability* to suboptimal conditions. Vulnerable periods occur at different times for different organs in the body.

2. Programming has *permanent effects* that alter the body's responses in later life and can modify susceptibility to disease.

3. Fetal development is *activity dependent*. Normal development is dependent on the baby's continuing normal activity in the womb.

4. Programming involves several different *structural changes* in important organs.

5. The *placenta plays a key role* in programming

6. *Compensation carries a price.* In an unfavorable environment, the developing baby makes attempts to compensate for deficiencies. However, the compensatory effort made by the baby often carries a price.

7. Attempts made after birth to *reverse the consequences* of programming may have their own unwanted consequences. Problems may arise when postnatal conditions prove to be other than those for which the fetus prepares.

8. *Fetuses react differently* to suboptimal conditions than do newborn babies or adults.

9. The effects of programming may *pass across generations* by mechanisms that do not involve changes in the genes.

10. Programming has *different effects in males and females.*

Prenatal programming occurs because our bodies act as societies of individual cells. Each organ is made up of different kinds of cells that have to live together and work with each other, just as people in a society must work together and tolerate each other if the society is to be harmonious. Some cells in the liver store glucose, while other nearby cells within the same organ have the opposite job, releasing glucose when food is not available. An imbalance of the proportions of these two types of cells can greatly alter the way you handle your snacks and possibly even your need to snack.

Cells also talk to each other, suggesting that they increase or decrease their activity for the common good. These conversations can be long-range, through hormones that surf around the bloodstream giving instructions. They may also be local, as when cells talk to the cells next door. Just like a society where everyone talks and no one listens, an imbalance between the numbers of cells giving instructions and the numbers of cells receiving instructions doesn't make for an efficient system.

A growing body, like a growing community of cells, faces special challenges. When all the community's needs are taken care of, it can grow normally, managing it's current needs while simultaneously planning for the future. But when the community faces shortages, it has to make choices that may haunt it in the future. For a community of people, these choices may include building fewer schools, or inviting in "dirty" industries and putting up with some pollution in exchange for jobs. If the developing baby in the womb has to focus too much on surviving in the short term, she or he may miss out on essential planning for the future.

Principle number one says that there are critical periods when certain parts of the fetus are vulnerable to stress from exposure to toxins or from lacking nutrients or oxygen. That is because, at many stages of fetal development, all cells have to make a fundamental choice between growth and specialization, just like people decide whether to take a job after high school and earn money (growth) or go on to college for further study (specialization).

Since the tasks cells must perform to grow and divide are very different from the things cells do to perform their special function in the mature body, growth and specialization are competing processes. Each cell must make a choice to divide or specialize at some point, and the process is irreversible. Some kinds of cells, like nerve cells, never divide again after they specialize. It is therefore very important that each developing cell makes the right decision at the right time. If the decision is made too early, an organ will end up with too few cells and thus be unequal to the challenges of later life. Cells that are forced to specialize too soon can be likened to bright and promising students who are forced by financial hardship to drop out of high school and begin working in a factory. With more educational development, these gifted young people might have continued developing their potential and become lawyers, engineers, artists, teachers, or professionals in some other more demanding career. Although workers do move up in the factory hierarchy, their futures are usually much more limited and they are more likely to be stuck with a certain role in life.

One of the most dramatic examples of a critical period is the development of sexual identity. There are a few specific days during development when exposure to testosterone will lead certain key areas of the brain to organize themselves in a male pattern. If the brain is not exposed to testosterone during this period, it will organize itself in a female pattern. Experiments done with developing hamsters and other rodents have shown that if female animals are exposed to the male hormone during this critical period, they will remain female physically but will act like males as adults. They will be sexually interested in other females and try to mount them. In effect, they have become males inside their heads, even though they are female externally.

As you might imagine, this kind of information has fed into a highly emotional debate about human sexual identity. Recent studies have shown that gay men tend to have older brothers, and that hormonal levels in the womb change for each successive son. Scientists have not yet determined whether there is a connection between the two findings. There are also other findings of common physical tendencies among gay men and women (such as differences in the lengths of the fingers) that must have their origin in environmental influences in the womb.

Principle number two says that programming has permanent effects. Growth and ultimate body size are good examples. When nutrition is deficient in the womb, cells divide less frequently, resulting in fewer cells in total in the baby's body. So a fetus that is subject to challenging conditions early in a pregnancy will be smaller overall, but all the body parts will be proportionately relatively normal. This form of growth retardation is called symmetrical growth retardation. When stress is making things difficult in the second and third trimester, however, the cellular community of the growing fetus prioritizes where nutrients will go. Since the brain is the organ most important for survival both inside and outside the womb, the lion's share of blood, nutrients, and oxygen gets routed to the head. Many other tissues suffer and don't grow as quickly as they should, and the newborn will have a head that is slightly large in comparison to the body. This pattern of growth is called asymmetrical growth.

Recently, researchers in Spain have documented a very interesting permanent effect of prenatal programming. They found not only that girls born with low birth weights have a smaller final stature when they have grown to their full height decades after birth, but also that many such girls will reach sexual maturity an average of 1.6 years earlier than girls born with normal birth weight. In some ways, this makes evolutionary sense. If a girl is being born into a world that is dangerous or is subject to famine, it would be wise for her to reproduce early, because there would be no guarantees that she would survive long enough to reproduce otherwise.

In both symmetrical growth retardation and asymmetrical growth retardation, the fetus is smaller than normal, but I must emphasize that a small fetus and newborn doesn't necessarily mean there has been growth retardation. Some babies are just genetically small, which is perfectly healthy. Other babies are small because the mother's small stature has kept their size down, which is also fine.

Principle number three says that fetal development is activity dependent. Another way to say this is that the fetus must "use it or lose it." The way fetal cells develop depends on how the fetus is using his body. He is able to swallow and suck at birth because he is practicing sucking and swallowing the amniotic fluid that bathes him in the womb. The nerve cells in the brain wire up correctly because signals are running through them, testing them, throughout the whole time they are developing.

Principle number four says that programming involves structural changes in the developing organs. One cause of these structural changes is the altered growth of blood vessels in a challenged fetus. If an organ develops too few blood vessels in utero, it will be harder for the body to increase the blood supply to that organ during times of need later in life. A paucity of blood vessels can affect how well organs like the liver, pancreas, kidneys, muscles, and heart carry out their jobs. If we liken your body to a productive field in a sunny but arid part of California, your blood vessels are like the irrigation system that brings water to growing plants. Without delivery of adequate amounts of water, no amount of California sun will grow the crop.

One interesting example of how programming can cause structural changes in the body is the development of fingerprint patterns. You (and many law enforcement personnel) might think that the formation of fingerprints is completely under genetic control. In fact, your fingerprints are not solely determined by your genes. The specific pattern of fingerprint ridges that form is determined in large part by the extent of swelling in the finger pads at the precise time when fingerprints are forming, around the tenth week of development. More swelling of the finger pads is more likely to produce a whorl pattern, less swelling is more likely to lead to a loop pattern or arch pattern. It happens that the arteries that serve the brain and head in the developing fetus are connected to the arms. When the prenatal environment is challenging, the fetus makes a priority of getting blood to the brain, which also happens to push more blood into the developing finger pads, causing them to swell. So when the fetus is short of oxygen for any prolonged period around the time the fingerprints are forming, and blood flow to the brain and head becomes a priority, there will be more whorls formed than flat arches.

Principle five is that the placenta plays a key role in programming. The placenta is a crucial organ in development, the bridge between you and your child. The baby's placenta not only acts as a gatekeeper for everything that comes into and leaves your baby, but it is an important hormone-producing organ that affects the way you and your child change physically and psychologically. Anything that important in development will by necessity be important in programming.

Principle six is that compensation for shortcomings during development carries a price. This is really the central point about how programming affects our health. For instance, consider how your baby's digestive tract and liver are affected by detrimental conditions. The organs that process digested food, like the gut and the liver, aren't really used much before birth, since the placenta takes care of the baby's need for food. The liver, a large organ in the upper part of the abdomen, has many important jobs, such as filtering out and processing waste toxins in the blood as well as storing glucose. Since most of your baby's waste can be processed by your liver, and glucose comes across the placenta all the time from your bloodstream, the baby's liver operates at less then 100 percent efficiency until after birth. At times when oxygen or nutrients are low, babies have a trick up their sleeves. They deprive the unessential organs of blood, sending the majority of their blood supply to the essential organs. But this ingenious act comes at a price. The rerouting of blood away from the liver forces the baby to grow a small liver. With nutrients and oxygen shunted away from the liver and gut during prenatal development, these vital organs are smaller at birth than they should be. Abdominal girth is also smaller in relation to the rest of the body.

The problem is that these deprived organs may never really fully catch up in growth. As a result, they continue to function less well than they should throughout life. Since the liver is central in regulating cholesterol, it is not too hard to see why growing a small liver, thereby reducing abdominal girth, will lead to higher cholesterol later in life. Hence the relationship that David Barker demonstrated between abdominal girth at birth and high levels of cholesterol in adult life. Inadequate growth of the liver and the pancreas, another key glucose-regulating organ, is likely responsible for the glucose control problems in later life, which can lead to adult-onset diabetes.

MAKING UP IS HARD TO DO

Principle number seven dictates that attempts to reverse programming after birth can have unwanted consequences. You might imagine that a baby whose growth was checked in the womb might be able to get back to normal if provided adequate nutrients, oxygen, and a low-stress environment after birth. But so many of the decisions that the cells make are irreversible. Low-birth-weight babies can get chubby again with plenty of food, but that is probably not the end of their problems. We know that animals that are presented with nutritional challenges in the womb and then a rich diet after birth actually have shorter lives than those that are kept thin on a low-calorie diet during life after birth. It is in this type of knowledge that we will find the key to helping people who were disadvantaged in their growth patterns before birth. As I will explain more fully in chapter 4, the fetus in the womb appears to sense what is happening in the world outside by virtue of the experiences he or she has in utero. Lion cubs in the womb on the Serengeti Plain whose mothers are short of food begin to specialize and prepare for a life in an outside world where food is in short supply. At such times they must grab hold of and store every calorie they will be lucky enough to find. Preparing for a life of nutritional shortages is an excellent survival skill, a survival trick that humans seem to have retained. However, in the human situation, babies born after experiencing adverse conditions in the womb generally grow up to find a world where food is plentiful, even if it is the wrong sort of food such as fat-laden fast food. So the bodies of these growth-retarded babies are not well adapted to the life of plenty that they meet after birth. Food is present in plenty and they are much more likely to become obese.

Providing a good womb environment for your baby may also help her to age more slowly. When pregnant rats are given a low-protein diet during pregnancy, their offspring show signs of accelerated aging in kidney cells. The DNA in these cells have shortened ends, called telomeres, which act as genetic timers for cells in the body. The telomeres in everyone's cells naturally get shorter with each cell division, and when they get too short the cells stop dividing; at that point the body stops being able to rejuvenate itself and starts to fall apart. Since the telomeres of those cells exposed to a low-protein diet in the womb start out shorter at birth, they may stop dividing sooner than they should. Control rats given a diet that promotes a good womb environment live longer and healthier lives than rats exposed to a low-protein prenatal diet.

Principle eight says that fetuses are special. They live by a different set of rules. They react differently to suboptimal conditions than do adults. There are many reasons that explain why developing fetal cells and organs are especially vulnerable to challenges. Growth is a demanding business. Growing cells have much greater energy and structural needs than cells that have completed growing and have settled down contentedly to perform their appointed functions. Growing cells need more oxygen, amino acids, vitamins, and glucose than those that are not trying to expand their activities. Little shortages can therefore damage or kill cells. Growing fetal cells are especially vulnerable to shortages in oxygen. Cells can store small amounts of nutrients and glucose, but they are quite unable to store oxygen. It is jokingly said that humans can only live without oxygen for minutes while they can live without water for days, food for weeks, and without ideas for a lifetime.

GENERATION UPON GENERATION

The idea that what happens to us at the beginning of life can reach out over decades and affect us to the very end is a revolutionary idea in itself. Principle nine reveals an even more amazing effect. The good or bad prenatal effects on one generation can carry down to the next generation, and even to subsequent generations. In other words, if your child is a girl, her own womb and other reproductive organs will feel the effects of the environment that you provided for her in your womb. If she is presented with challenges to growth, her body may be smaller than it should be, she may have difficulty keeping her blood sugar in balance, her kidneys may not function correctly, her response to stress may be higher. When she grows up and becomes pregnant herself, these factors may affect the quality of the womb environment that she is able to provide for her child. As we will see in the following chapters, problems in your own pregnancy can help create the

same conditions in your daughter's pregnancy. Blood sugar problems during your pregnancy can damage the blood sugar regulation in your daughter, leading to the same problems during her
pregnancy, and so on through generations. The good side of this
story is that healthy pregnancies can likewise improve the health
and abilities of our children and grandchildren.

The last principle tells us that programming is different in
male and female fetuses. Boys and girls grow in slightly different
hormonal environments. Their cells have slightly different
choices to make. I'm not just referring to the growth of the reproductive organs. It is clear that the brains of men and women have
certain very distinct differences in structure. Women have more
gray matter (the areas with nerve cells) while men have more
white matter, the connections between nerve cells. These differences are subtle but may play a major role in behavioral and other
functional differences between men and women. It is not surprising therefore that the effects of a suboptimal environment in the
womb may act differently on men and women. Depression, for
example, is more common in women than in men, and part of this
difference may well be due to the different ways in which the
brains of male and female babies react to stress and other adverse
conditions in the womb.

MAKING A DIFFERENCE

Fortunately, understanding these principles can enable you, your
spouse, and your health care provider to take the appropriate action
to minimize adverse effects. On a day-to-day level, that means that
you are entitled to pamper yourself, and your family should pamper

you. Everyone around you needs to know the importance of providing a good supportive environment at this time.

On a national level, prenatal health care should move up on the national agenda. If we as a nation would put more dollars into prenatal care, we would see great improvements in the national health. In fact, the importance of prenatal health care for adult health may explain a fascinating and perplexing mystery that has puzzled epidemiologists, cardiologists, and nutritionists for decades. This mystery is called the French Paradox, and it goes like this: since eating a lot of fatty foods, especially animal fats, is considered, according to conventional wisdom, to lead to high levels of bad cholesterol and high blood pressure, the French should have a high rate of heart disease. The French, after all, slather butter on and in everything, and eat plenty of red meat, duck confit, pâté, and other rich foods. According to all the current dogmas, the French should be dropping like flies from heart disease at an early age. They aren't. Their health is actually better than we enjoy in the United States. Explanations have included the presence of a Mediterranean diet with lots of olive oil, or the health benefits of moderate red-wine drinking, or simply (a Gallic shrug here) the unexplainable superiority of everything French. In fact, the answer to the French Paradox may lie not in current eating habits, but in far-sighted French policies for the care of pregnant women introduced over a century ago. In the late nineteenth century the French were the first in Europe to begin a program of caring for pregnant women in order to offset the adverse effects on their pregnancies of strenuous physical work in the sweatshops and laundries of Paris. The subsequent generations of healthy pregnancies and good prenatal practices may be the real secret to the French Paradox, the real reason that the French can enjoy life to the fullest in the best of health.

2

The First and Best Home for Your Baby: The Optimal Womb Environment

I have to admit that I am addicted to curries. Curried rice, curried lamb, curried beef, green curries, red curries, Indian curries, Thai curries—I love them all. I used to joke that the reason I am so partial to curry is that I started tasting it in my mother's womb before my birth in Sri Lanka, where my father was a civil engineer. My family moved back to England when I was young and my passion for curry didn't begin until I was much older, so I thought there was nothing more to my curry-in-the-womb theory than idle banter. But as I will show you, more recent research has shown there is probably something to the story.

The world of the womb affects the fetus far more than we ever imagined, and not just in terms of lifelong health. As you will see, every step of normal development is a cooperative act between mother and child. Development requires an active give-and-take between the two that affects the baby's long-term thoughts and senses.

• • •

WINDOW ON THE WOMB

The womb is a microcosm in the truest sense of the word. To the fetus the womb is his whole world, a universe in miniature. Living in warm, fluid darkness, the fetus is supported by the gentle bands of his mother's uterine muscle, rocked this way and that by her motion. Your first ultrasound pictures of your baby show him on his journey to a new world. He is a voyager in inner space, eyes open and innocent, dreaming fitfully, an astronaut tied to his mother ship by a thin, transparent, umbilical lifeline.

For a long time, this comforting picture is about all we had of life in the womb. We envisioned the fetus drifting quietly, with an occasional kick, passively waiting to be built according to his genetic blueprint. We thought that this state of affairs continued until he had grown too large for his mother's body, whereupon she expelled him from his twenty-four-hour-a-day Jacuzzi into the cold, bright, world outside the womb.

New imaging technologies and observation of developing animals have changed this vision—dramatically turning it on its head. By using ultrasound, you now get snapshots, and even video movies, of your baby before birth. Scientists can find out even more. By closely studying patterns of movement and monitoring subtle hormonal and neural signals, they know that your baby is more than just physically active—he is communicating constantly with you, receiving signals about the world outside the womb that help him learn and practice the whole range of skills he will need later in life. Through this ongoing chemical and physical conversation, your body is instructing your child about things he will need to know to become a self-sufficient person, coaching him to prepare for a healthy life. What is happening to you in your daily life tells him about the challenges he will face when you are no longer surrounding him with a protective envelope.

Not only does the picture of your baby as a passive participant not do him justice, but in many ways, it is your baby who is controlling your body. He instructs you to make more blood, to eat different foods, to loosen joints, and put on fat. He even sends you signals when he's ready to be born. The science of fetal physiology began in Cambridge, England, fifty years ago. Laboratories throughout the world have demonstrated clearly that the set of biological rules by which the fetus lives are very different from the rules you and I have to obey. Biologically, your baby is cleverer than he will be at any other time in his life. He has to live in one world according to one set of rules while at the same time developing the systems he will need to live in another world with a completely different set of rules. He needs all the help he can get from you.

We are rapidly learning what you and every pregnant woman should try to do as you strive to create the best possible environment for your baby's growth and development in the womb. The right kind of environment will allow you to offer the right kinds of messages to your child so that, at birth, he is already well along the most healthy developmental path. The right messages are those that tell your developing baby to be at ease and not to worry. These messages tell him that the world outside is a safe place, with plenty of food and comfort. They allow him to grow quickly to his full potential, just as a five-year-old—comforted and shielded by his parents from the harsh realities of adult life— grows up to be psychologically stronger than he would if he were constantly exposed to troubling events.

MAKING A HOME IN THE WOMB

Since the program that will unfold in these pages is founded in knowledge, I would like to spend a little time talking about how the fetus grows and how he communicates with and learns from you, his mother.

Let's go back to your baby's very first beginnings. The fertilized egg does not grow until it links up with the lining of your womb. As the newly fertilized egg drifts down from the ovaries toward the womb, it divides again and again, but its overall size is not increasing. On the contrary, since the embryo is not receiving any additional food, the cluster of cells drifting down the fallopian tubes remains slightly smaller than the period at the end of this sentence. Once this pinpoint cluster snuggles into the wall of your womb, however, it begins sending out commands that reorganize the very structure and function of the tissues that surround it and are its immediate environment. One of the lines of communication is through your blood supply, a large part of which is rerouted to bring nutrition to the womb. From that moment to the end of pregnancy and beyond, communication between you—the mother—and the embryo* (and later the fetus and the infant) is inextricably linked to healthy growth and development.

It is fantastic to realize that the first one to communicate with the baby is you. It is even more fantastic to know that, even before you start sending signals to your baby, he starts sending them to you. Shortly after implantation in the womb, the embryo starts releasing chemical messages that open the lines of communica-

---

Embryo is the name used from implantation until eight weeks of gestation, and *fetus* is the tem we use thereafter.

tion between the two of you. A hormone called chorionic gonadotropin, enters your blood and travels through the bloodstream to the ovaries. There, it tells the empty follicle that held the egg not to die but to keep producing other hormones to maintain the pregnancy. Without chorionic gonadotropin, this empty follicle would stop producing these hormones after two weeks, initiating menstruation. Chorionic gonadotropin is the hormone that is detected in a pregnancy test.

Growth factors produced by your developing baby promote the growth of blood vessels and capillaries from both your own tissues and his: these will form the placenta, allowing the exchange of oxygen and nutrients for carbon dioxide and waste products. So the rapidly growing placental network is an essential organ that is part your baby's tissue and part your own. The placenta will sustain your baby throughout nine months, acting as a lung, a kidney, and a digestive system before his own are working, but it is eventually thrown away.

Your baby's blood supply is directed to the placenta through arteries in the umbilical cord; it flows into myriad tiny capillaries in the placenta and returns—now rich in oxygen and nutrients that you have provided—via the umbilical vein to the fetus's body. Although your blood circulates very close to your baby's in the placenta, they never mix. The two bloodstreams are separated by a thin layer of tissue, across which everything your baby needs is transported. The umbilical cord is your baby's supply line, and the placenta is the transport depot where the supplies are loaded. Growing a good placenta, and the proper function of this intimate linkage between you and your child's blood supply, is the foundation of good fetal growth and your child's future health. Thus it is very important that you eat well, rest well, and do the right things from the very beginning of pregnancy. Although you are not growing much in this phase, your baby is setting the stage for a healthy pregnancy.

The placenta also produces an extraordinary range of hormones. Hormones are chemical messages that are produced and secreted into the blood by one cell and then travel around the body to tell other cells what to do. They are the language in which cells talk to each other. When it comes to producing hormones, the complexity of the placenta matches the complexity of all other hormone-producing glands in your body put together. The placenta you share with your baby is probably the most talkative and bossy biological organ there is. The placenta sends firm instructions to you as well as to the baby. In early pregnancy it can be considered almost more important than the cells that make up the growing embryo. One of the most notable hormones that the placenta produces is placental lactogen. This hormone is responsible for the swelling and tenderness of your breasts at the beginning of pregnancy, and for preparing them to produce milk when the pregnancy is over.

Placental lactogen is just one of the many hormonal signals that the baby and placenta send to reshape your body and bend it to the baby's needs. Between the two of you, the baby gets first priority—and making you feel terrible is fair game as long as it doesn't put your health in danger. So one of the things that these changing hormonal signals do is make you feel very ill. Morning sickness is the name given to the nausea and vomiting that can afflict you any time of day, especially in the early stages of pregnancy. Nearly nine out of ten women feel morning sickness at some time during pregnancy, and 15 percent feel morning sickness all the way through. Men usually don't understand just how terrible it is to feel sick all the time, how draining and dispiriting. "It's like being seasick . . . for months," is how one woman described it to me.

Although generally the worst thing about morning sickness is that it is draining and dispiriting, occasionally it can be dangerous. When I was practicing obstetrics in London, I remember well caring for several worn, pallid women who were admitted to the hospital because they were throwing up almost constantly. These pitiable women couldn't keep anything down—not food, and very often not even water. I worked to keep them hydrated with intravenous fluids and tried to help them keep food down. Luckily, extreme cases of morning sickness are very rare. However, if you are worried that you are one of these rare cases, you should discuss your nausea with your obstetrician to ensure that it lies within normal, probably protective, limits.

In times of extreme sickness it may also be helpful to know that in almost all cases, morning sickness is not harmful to your pregnancy. In a recent extensive review of the scientific literature on morning sickness, a friend of mine at Cornell University, Paul Sherman, concluded that there is no evidence that morning sickness leads to miscarriage or birth defects. Sherman found quite the opposite: moderate morning sickness appears to be protective. The question is, why?

There are two theories. One theory comes from an American researcher, Margie Profet, whose training as a biological scientist inclined her to look at behaviors in terms of their evolutionary advantage. The fact that morning sickness is such a widespread phenomenon made her wonder if there was an advantage to morning sickness that was not at first apparent. Profet began looking at the kinds of foods that pregnant women avoided most on account of nausea—garlic, broccoli and other members of the cabbage family, spices, foods that may be aged or old—and found that these kinds of foods are most likely to contain toxins that could threaten an embryo. The experience of nausea also might confer other benefits on pregnancy since nausea promotes a varied diet: after eating something while nauseous most people tend not to want that same food again. Profet has taken some criticism for seeming to say that "vegetables are bad," but what she is in fact saying is that some vegetables protect themselves against bacteria and other plants by creating toxins. So don't feel "crazy" for shunning foods that you loved only a few months ago. Don't worry if you are trying to eat more vegetables but can't stand certain types. Your nausea is probably only your baby making you suffer in order to protect himself. You'll be doing the same thing for your child as he grows up!

Many leading endocrinologists feel that sickness is also a side effect of the body's attempt to alter digestion and metabolism to favor energy storage. This is the theory put forward by a Swedish scientist, Kirsten Uvnas-Moberg, who noticed that the time of greatest sickness, when you would expect someone to be losing weight, are when women put pounds on. Uvnas-Moberg has done research showing that the biochemistry of the body changes during pregnancy in order to favor energy storage over energy use. Nausea, heartburn, fatigue, and slower movement of food through your digestive system are the result of the digestive system trying to get every ounce of energy out of food and then storing that energy in fat. This leaves working cells short of energy. These changes in digestion are created by an increase in the production of certain signaling molecules that normally elevate levels of insulin in the blood and make people feel tired after meals even when they are not pregnant. In pregnancy, when more of these signaling molecules are produced, you may feel tired all the time, not just after meals, and the changes in blood sugar produced by increases in insulin can make you feel shaky and weak.

Weakness, shakiness, and nausea can also be the result of

changes in your blood volume, one of the major adjustments you will make to being pregnant. As soon as your pregnancy begins, your baby signals to you that you need to produce more blood to carry the oxygen and food he needs to the placenta. But it takes a few weeks to make the necessary new blood, and so for a time your body may feel that it has too little blood for the tasks at hand. This is not really a shortage yet, but if you do not react by producing more blood, the baby will be shortchanged later when he really starts growing. Again, these preparations show how marvelously well coordinated are the early events of pregnancy. Slight feelings of weakness, nausea, and sudden coldness in the limbs are experiences that every pregnant woman has encountered. This might seem like a terrible time to put more demands on the blood supply by exercising, but in later chapters I will tell you why the appropriate amount of exercise may actually help along these changes in your blood.

ALTERING THE GENETIC BLUEPRINTS

I hope I have now convinced you that, contrary to the traditional view, the embryo or fetus is not playing a passive role in the pregnancy. Far from it; your baby is driving the experience, steering your thoughts, emotions, and bodily sensations. But the little taskmaster in the womb is also listening to news about the world outside, like a listening post in distant space seeking information about life on other planets. He seeks information because he needs to get ready for the world outside the womb. He does this by modifying the action of his own genes.

Most people can be forgiven for thinking that the way in which babies develop is completely predetermined in an unalterable way set by their genes. We have what I call "gene myopia" in our society: a belief that it is the genes alone that determine our health and well-being throughout life. We are constantly hearing exciting news of scientists finding the gene for this or that trait— aging, depression, left-handedness. The mistaken interpretation of these findings is that these complex traits are fixed by the genes, rather than just influenced by the genes. We now know that although our genes are indeed fixed at the moment of conception, the more complex the trait involved, the more the environment in the womb affects what our bodies do with those genes. Information from the environment around the fetus, in other words, helps determine which genes are switched on, which are switched off, and when these alterations in gene activity occur. We can see that genes don't determine exactly how babies develop by looking at identical twins (who form by the splitting of one fertilized egg and therefore share the same genes) and fraternal twins (formed from two separate fertilized eggs and sharing only some genes, like any siblings). When researchers started testing the gene sequences of twins they were using in research, they were occasionally surprised to find that twins who looked fairly different, and therefore were thought to be fraternal twins, had the same genes, and were in fact identical twins. Somehow, through their placement in the womb, or simply by the different decisions that key cells made early on, one twin had developed along a significantly different path, even though each cell was driven by an identical set of genes. These discoveries have led me to make the somewhat controversial judgment that there are no such people as *identical* twins. They may come from the same egg and have the same genes, but they pass their critical biological milestones under different conditions. Twins in the womb crowd each other for space and often compete for their mother's resources. If they share a placenta, that competition is more fierce. Such competition leads to differences in the kind of environment each twin is exposed to.

The reason that cells create their own variations on the genetic plan is that it helps us survive. If a baby can change the pattern of development to better suit life in the womb and the conditions he will be likely to meet after birth, he is going to be better prepared than babies who can't. We shall see that the fetus is clearly listening for signs of two of the most important challenges to be faced during life after birth: hunger and stress.

Your baby can monitor the amount of food he is receiving across the placenta. If there is a shortage, he begins to learn about hunger and prepare for potential food shortages both now and later. If there is not much food available in the world outside, the fetus will be better off if his body is wired to store away all available energy in the form of fat. These profound and often permanent changes in the appetite and the way we handle food are best made when the body is just forming.

Your baby also monitors your stress level, constantly measuring the amount of stress hormones that come across the placenta into his blood. The more of your stress hormone that appears in his blood, the more the world outside the womb seems a stressful and threatening place. In this case, he alters his development so that he will react effectively to threatening situations. Brain cells adapt so that they are hypersensitive to stress. On the other hand, when the signals are right, the genetic program expresses itself fully and your baby develops steadily along an optimal track. It appears that, as in so many other things in life, the secret of fetal development is to enable your baby to match his development with his own personal characteristics, not those imposed on him by an adverse environment.

BENEFICIAL LEARNING

On the more positive side are the multitude of exciting and beneficial interactions going on. The fetus receives many positive cues
from you that are necessary to healthy development. Your baby
translates your daily bodily rhythms of sleep and wakefulness, eating and drinking, into his own bodily rhythms. Even the rhythms
of his heart rate become locked into yours, truly a heart-to-heart
connection. Nerve circuits, and other control circuits in the body,
practice in order to wire up correctly. Your baby is making breathing movements in the womb that will prepare his lungs to breathe
after birth. By taking information from the womb and the world
beyond and building that information into the expanding and
developing circuitry of his brain and body, your baby is learning
and developing. He does so by monitoring

Sound: Hearing is a critical skill after birth, and preparation to
hear what is going on in the world has to begin long before delivery. Luckily, the womb is awash with sound that can help the brain
wire up its sound circuits. There are the gurgles and squelches
that your stomach makes, the regular beat of your heart, the
sound of your voice. The baby is scanning and picking up sounds
like a ham radio operator. Researchers have shown that newborns
prefer to hear the sound of their own mother's voice over the
voices of others, because her voice has been the clearest and most
familiar sound all through pregnancy. Newborns are also much
more attentive when they hear songs that were played regularly
while they were in the womb, or hear stories that their mothers
regularly read while pregnant.

Touch: The gentle bump and caress of the womb's wall is a critical part of learning to respond to touch. As a mother moves, her
motion jostles and nudges the fetus, who jostles and kicks back. In

addition, throughout pregnancy a mother's womb is contracting gently every once in a while. Many women feel these contractions. These tightenings, often lasting several minutes, are called "contractures" or "Braxton-Hicks contractions" to distinguish them from the contractions of labor. The contractures act as gentle uterine hugs (as Dr. Thomas Kirschbaum from the University of Alabama calls them) that stimulate the baby's nervous system to begin working properly. They help integrate sensory and motor neurons in the body and brain.

Taste: Even while in the womb, your baby is swallowing fluid and her taste buds are learning to recognize tastes. Scientists find that if they inject a sweet fluid into the amniotic fluid, the fetus seems to like the flavor and will increase the amount of fluid she swallows. Some recent research suggests that taste preferences for garlic, spices, and other foods may be set and programmed before birth. The idea is that these flavors filter into the amniotic fluid from the mother's blood. Since your baby is swallowing and tasting amniotic fluid, she is getting used to certain flavors and forming preferences for those flavors before birth.

Julie Menella, a researcher at the Monell Chemical Senses Center in Philadelphia, followed women who drank carrot juice regularly while pregnant, and then offered their babies cereals with or without carrot juice six months after their birth. The babies exposed to carrot flavor in the womb (and those in a separate group who tasted it while breast-feeding) were much more partial to the cereal with carrot than cereal that did not contain the carrot flavor.

So when you expose your child in the womb to various flavors, you are teaching her about the kinds of foods that you eat, the foods that she will probably eat as she grows up. This culinary preparation helps teach her about good eating, but also helps her fit into the society she is about to enter. Hence I was not surprised to find out that my mother loved curry dinners on those sultry evenings in Sri Lanka so many years ago.

Biorhythms: Certain chemical compounds have distinct twenty-four-hour rhythms in your blood even when you are not pregnant. Stress hormones, kidney hormones, and melatonin naturally rise and fall over the course of the day and night. Most of these hormone signals pass across the placenta to some extent at least. Researchers have shown that fetuses are aware of these changes. They have determined that developing animals use these changing signals to synchronize their own activity cycles to that of the world outside the womb, so they can be better in synch when they are born. Human fetuses also seem to use these signals to learn about the cycles of the world outside and to try to adapt their own cycles to them. So living well during pregnancy is likely to pay dividends in helping your newborn baby to have good sleep patterns after birth.

OPTIMAL, NOT PERFECT

Improving the quality of the environment in the womb and communications between you and your child does not mean trying to do the impossible, namely to create the perfect womb. Therein lies a scenario for needless worry and guilt. There is no such thing as a perfect womb environment, just as there is no such thing as a perfect parent. Life is full of ups and downs. We can't always be in a calm state of mind when caring for an infant, although we try our best. You can't expect yourself to always be calm when pregnant, or eat only prescribed foods.

The goal of your program should be to give your child the healthiest signals possible, using our new understanding of prenatal programming. What you should strive for is the optimal achievable environment, one that is the best possible under existing conditions—which are the stressful and difficult realities of being a pregnant woman in the modern world.

3

Nutrition in the Womb: We Are What Our Mothers Ate

When expectant couples ask me advice about providing the best environment in the womb, I used to first ask them if they were eating well. But recently I realized that this was entirely the wrong question. For one woman, eating "well" meant eating a fair bit of food every day. One man seemed to think that a "balanced" diet merely meant eating regularly. Another woman thought "eating well" meant eating healthy food. No problem there, except that for her, healthy food consisted almost completely of vegetables and fruits, since that was what she ate most of the time to lose weight. Now I simply ask about what they ate that day, and then take it from there.

My first questions are about nutrition because diet has such an important effect on the developing baby, and eating well is something you can work on right away. So much of what happens during pregnancy is not completely under your control. Your baby inside has a way of taking over your body and mind, making you feel nauseous, moody, bloated, and clumsy. Baby gets bigger and bigger, whether you want him to or not, advancing you toward labor without pause or time-out. What you can control is your diet. This is good, because what you eat—and therefore what your baby eats—is critical to your child's development. This is further illustrated by astounding recent discoveries:

- Your nutritional state before conception is as important, perhaps more important, than your nutritional state during pregnancy.

- The health and nutritional state of your body—the baby's home—is the single most important factor in the baby's growth.

- In cases in which a fetus is extremely nutritionally deprived in the womb, it may take a few generations to undo the damage.

- The effects of poor nutrition vary according to the trimester in which the problems occur.

For thousands of years, the human species lived on a diet that consisted mostly of vegetables and grains, with some meat and fats and almost no sugar. In the last century, the developed world has in general become much more affluent, able to afford more fatty meats (even prizing the "marbling" of fat in steaks) and sweets. The less affluent tend to sustain themselves on large portions of starches like flour and potatoes. Almost everyone eats fewer vegetables. In the past few decades, our dietary situation has become even more precarious. Everywhere you look, there are tempting snack foods with empty calories and few essential nutrients. Every commercial district carries at least one fast-food restaurant offering high-fat french fries and hamburgers. By eating just one "supersized" meal, an adult can consume her caloric requirements for the whole day. The frozen-food section of grocery stores expands into more and more aisles as people living hectic lives are ever more dependent on prepared frozen foods. These foods are seductive not only because of their rich, engineered flavors and their ease of preparation, but because they are often even cheaper than buying groceries and fixing food yourself. These temptations hardly existed forty years ago, and people thought of the potato chips or other snacks that were available as "kids' stuff" that adults never took seriously. It is now much, much easier to succumb to temptation and not get a balanced diet of freshly prepared foods.

We see the effects of society's nutritional decisions all around us. For instance, we are currently in the midst of epidemics of obesity and diabetes, according to public health officials. Statistics show that both conditions are about double what they were just ten years ago, reflecting changes in eating habits thirty to forty years ago. In the United States, almost one in five people is obese. Doctors are seeing dramatic increases in the number of people developing type II diabetes, which is also known as adult-onset diabetes. Type II diabetes is even appearing increasingly among children, who have not historically been affected. This increase cannot be due to genetics—diabetes is rising too rapidly. Researchers consequently focus on lifestyle and diet as the factors responsible for this astonishing deterioration in public health. For decades, scientists have been studying the effect of adult diet on health and longevity, and have also been looking more closely at how diet during childhood and adolescence predetermines health and longevity as an adult. But, as we saw in Barker's work, that doesn't explain everything. Nutritional scientists are finding that obesity, heart disease, and many other diseases are programmed by the nutritional environment in the womb.

PRENATAL EATING HABITS

In a sense, eating habits start before birth. The fact is that even as the child is forming, his body is developing "eating habits," habits determining how the cells and organs in the body will process food and how they will incorporate nutritional resources into their structure. These cellular habits have a direct and powerful influence on how well the body is able to handle threats and insults during life after birth. Since prenatal nutrition can affect how your child processes food in later life, cellular habits formed in the womb can also affect actual eating habits later on. It seems there may even be specific "critical periods" of development in the womb, when the types and amounts of nutrients available determine the body's nutritional response thereafter. We do know that over the whole course of prenatal development the kind of food your child gets before birth can affect his lifelong

- Weight
- Blood pressure
- Blood sugar profile
- Cardiovascular health

We know these associations between nutrition during pregnancy and long-term health from observing the most nutritionally stressed children in studies of starved or diabetic mothers. But we also know that you don't have to be poor, sickly, or starving to stress your child nutritionally. Many women simply don't know the rules for healthy eating during pregnancy. They either don't have a working knowledge of the guidelines for healthy eating generally or don't know about the special nutritional requirements of pregnancy. Or a crisis sweeps away any ability to choose what you eat, as happened in Holland over fifty years ago.

HISTORY'S CRUEL EXPERIMENT:
THE DUTCH HUNGER WINTER

Much of what we know about good and bad nutrition during pregnancy comes from animal studies, the most precise leg of the "triumvirate of evidence" that supports prenatal programming. But occasionally history records natural experiments in human suffering, from which we are able to learn a great deal. Many of these disasters occur in developing countries where food shortages and famine can happen every few years. In these instances people might experience food shortages many times in their lives—during their time in the womb and then a few more times during childhood and adolescence. But one of these horrible events is particularly illuminating because the famine happened for a very distinct, limited period, in a well-off, first-world country. So we can start with a piece of human history from the end of the Second World War.

Throughout history, Holland has found itself at the crossroads of armies marching across Europe. The closing years of World War II were no exception. In September 1944, British and American paratroopers dropped into Holland in an attempt to seize a key bridge over the Rhine River. The idea was that the paratroops would take the bridge and hold it until the Allied armies could advance to support them. As depicted in the movie *A Bridge Too Far*, a German spy passed on the information about the airdrop and there were several panzer divisions waiting for the lightly armed airborne troops. The attempt failed and the parachute troops were captured or killed.

The Allied soldiers were not the only ones to suffer. The Dutch government in exile had asked the Dutch populace to rise up and harass the Germans as the operation began. After the operation failed, the Germans responded savagely by severely restricting the food supply to western Holland. Before the restriction on food supply, the average person consumed 1,500 calories per day. After the Germans retaliated, the daily diet was cut to 750 calories per day. At the height of the very harsh winter of 1944–1945, some people were able to scrape together only 450 calories per day. A whole day's ration might consist of a couple slices of bread, a turnip, and one or two small potatoes. Many Dutch resorted to eating their famous flower bulbs. Those who lived in the cities suffered more than those in the countryside, where hidden food stashes were more readily available. It is estimated that twenty thousand people died before the Allied troops liberated the region in early May 1945.

Considering the fact that modern dietary guidelines suggest that pregnant women consume about 2,500 calories per day during pregnancy, a diet providing 450 to 750 calories per day was terribly hard on growing babies still in the womb. Normally women gain about 25 pounds during pregnancy, but some women pregnant during this period gained as little as 4.4 pounds. The government gave extra ration tickets to pregnant women, but because people were blissfully unaware of fetal programming, these women's families often made the pregnant women who received the supplements share the extra food among their brothers and sisters. It is easier to see the needs of starving children standing before you than those of children not yet born.

I spoke to one of the survivors of the Hongerwinter (as the

Dutch call the period), a Mrs. Bakker, on a visit to Holland in 1996. Mrs. Bakker was one of the eldest of fifteen children in a Jewish family that lived in a farming community. Being Jewish, the family was forced to stay hidden and seldom went outside the small rooms where they lived. Mrs. Bakker became pregnant around the time the German reprisals began, and as the cold months dragged on she moved around very little to conserve calories. Since the period of starvation lasted almost nine months, Mrs. Bakker's baby daughter spent the whole of her time in the womb undernourished. As soon as the Allies liberated Holland in May 1945, they directed tremendous amounts of food into the area, and the average caloric consumption leaped to 2,000 calories per day. Nutritionally speaking, the change was all the more dramatic because the food they brought in included a lot of foods that were high in sugar and fat, such as candy bars and sausages.

A Harvard University Medical School pediatrician named Clement Smith saw a tremendous opportunity to investigate the effects of starvation on the outcome of pregnancy. He rushed to the area to gather birth data of babies born before, during, and after the Hunger Winter. His study provides some of the best available human data on this subject, and the results are highly interesting.

The first thing that Smith established is that undernutrition during the Hunger Winter was generally responsible for a significant drop in birth weight and a small drop in birth length. What also became apparent was that the precise time at which the undernutrition occurred in the pregnancy was very important in determining how babies fared. For instance, some women were already pregnant when the food shortages began. Although they gave birth during the Hunger Winter, their babies were only undernourished in the last weeks of their life in the womb. Other women, including Mrs. Bakker, conceived early in the period of food shortage and gave birth just as it ended in May 1945. Measurements of babies' length, girth, weight, and head circumference show that the greatest effects of undernutrition occur in the last trimester, when the baby should be growing the fastest.

One might think that babies conceived in March 1945 and therefore undernourished during the first few months of development during life in the womb might also be smaller, but this turned out to be untrue. These babies were actually fairly large, and there is a very clear and interesting scientific explanation for that. During the first few months in the womb, the Dutch babies were not actually gaining that much mass, and instead spent most of their energy growing their placentas. In response to the nutritional shortage during the first trimester, each fetus sent signals to her placenta instructing this vital supply line to grow larger to try to compensate for the shortage. When normal nutrition resumed in the second trimester, the larger placenta was able to make the most of the improved supply of nutrients and passed along more food to the fetus. By far the most badly affected babies were those who were conceived near the beginning of the Hunger Winter and lived through the whole of the period of deprivation while in the womb. These babies were shortest, weighed the least, and had the smallest heads. They also had the most health problems, and mortality levels that were ten times higher than normal.

Body shape was also affected by the precise timing of the malnutrition. When fetuses were starved in the first trimester, their developing bodies conserved energy by creating fewer cells, resulting in smaller bodies overall. Everything about the baby was smaller, and as a result their growth retardation was symmetrically distributed throughout the body. Undernutrition in the second half of pregnancy imposes a different pattern of growth retardation. By the middle of pregnancy, the fetus is well in control of his physiology and makes clever adaptations to channel most of the nutrients to the most important organ, the brain. Brain development is favored at the expense of organs that are not immediately important to survival in the womb, like the skin, bones, liver, and gut—all of which won't be utilized fully until after birth. When babies go through this, as many of the Dutch babies did, they are asymmetrically-growth retarded—born long and thin with heads that are slightly large for the size of the body.

Children of the Dutch Hunger Winter are now over fifty-six years old and still show the effects of their prenatal trial. Children who were nutritionally deprived in the womb were born shorter, and they remained shorter throughout life because their bodies had fewer cells than people who were not so starved. The long-term effects of starvation also were different depending on when in pregnancy the undernutrition had taken place. Scientists studied 300,000 Dutch army recruits who were in the womb during the famine, and found that some were more likely to be obese than others were. Those who were subjected to undernutrition in the first half of pregnancy were more likely to be obese at nineteen years of age, while those who were born a few months into the famine and only starved in the last trimester of pregnancy were less likely to be obese. One theory to explain these differences suggests that the centers of the brain responsible for appetite seem to have been affected differently depending on whether the nutritional supplies to the developing fetus were lacking in early or late pregnancy. Again, according to the first principle of fetal programming, the timing of the insult has different effects on different fetal organs.

MODERN EFFECTS OF PAST POOR NUTRITION

The nutritional problems caused by war are an extreme case, but everyday examples of poor nutrition can also have extreme effects on your children. There is increasing evidence that poor nutrition before birth can raise your child's risk of obesity and diabetes much later in life. For instance, the rising rate of obesity today may reflect the consequences of poor nutrition during pregnancy in some women as recently as twenty or thirty years ago.

The reason that obesity is connected to poor prenatal nutrition is that pregnant mothers who eat poorly or try to keep their weight unnaturally low during pregnancy will be programming their children's bodies to expect shortages of food after they are born and to be better at hoarding calories. Children who learn these tricks of metabolism in utero are more likely to be overweight or obese when they grow up. Some researchers refer to these individuals as having a *thrifty* metabolism.

A modern example of such programming is found among Ethiopian Jews, called the Falasha, who were offered a new home in Israel during times of strife in their native land. Like other Ethiopians, the Falasha had experienced chronic food shortages for generations, and most of them had undoubtedly been exposed to food shortages while still in their mother's wombs. After they were transported to Israel, a land of relative plenty, many Falasha soon had a problem that they had never encountered before: obesity. As a result of relatively poor nutrition in the womb, their bodies were set to make the most of the calories that came along. That strategy would have worked well had they remained in Ethiopia, but it had terrible consequences in their new land.

The connection between poor prenatal nutrition and diabetes is a little more complicated. Like obesity, diabetes may also be on the rise because of metabolic programming during fetal life. Animal studies have shown that poor nutrition in the womb alters the growth of the pancreas and the function of insulin. As a result, the very way that the body processes, stores, and utilizes energy is altered from birth, though it may take years for the effects to surface.

Diabetes can be thought of as a "cash flow" disease. One of the body's basic sources of energy is glucose, which is not just a component of sugar but is also the main component of starches like flour, rice, and potatoes. Glucose is the main form of "cash" that makes the body run. When you eat, glucose leaves the gut and enters the bloodstream. In response to rising blood glucose, the pancreas releases insulin. Insulin allows glucose to leave the bloodstream and enter muscle and fat cells where it can be used. In diabetes, the insulin doesn't work as it should, either because the pancreas doesn't release enough of it or because cells in the body on which it acts are deaf to its signals. Fat in the bloodstream is one major way to decrease the effectiveness of insulin. So it is not surprising that diabetes and obesity are linked, and that it pays not to get overweight. With insufficient insulin, glucose builds up in the blood while cells bathed in that glucose-rich blood are starving from lack of energy. Diabetes has sometimes been called "starvation in the midst of plenty." These starved cells are forced to consume their own fats and proteins as an alternate source of energy. When fats are mobilized and circulate in larger amounts, they may deposit in the wrong places. This combination of using up the body's cells and abnormal fat deposition can weaken cells and lead to many other sorts of disorders in the eyes, heart, and many other parts of the body.

By far the most common form of diabetes is type II, or non-insulin-dependent diabetes mellitus (NIDDM). NIDDM accounts for 90 percent of all cases of diabetes. It does not have a clear genetic cause. People with NIDDM usually become diabetic when middle-aged or elderly. Type I diabetes (insulin-dependant diabetes mellitus, or IDDM) is the more rare and dramatic form of the disease, which usually strikes before age thirty and requires daily injections of insulin.

People who get NIDDM have relatively normal blood sugar levels when young, but as the years go by, either their cells become less and less sensitive to insulin, or the pancreas produces less insulin, or both. This "insulin resistance syndrome" leads to small blips in blood sugar—hyperglycemia—after meals, which can damage the insulin-producing cells in the pancreas. When this process is repeated month after month, year after year, the insulin insensitivity increases. Further spikes in blood sugar cause more damage to the body.

The current evidence points to poor prenatal nutrition as one major cause of NIDDM. A recent Danish study on identical twins—twins with the same genes—supplied strong evidence that NIDDM is more related to impaired growth before birth than to any specific gene. The study found that when one twin became diabetic later in life but the other did not, the twin that got diabetes was more likely to be the one with the lower birth weight— the one that got the shorter end of the stick in the competition with her sibling during life in the womb.

If you are pregnant and have hyperglycemia or full-blown diabetes, it will definitely increase your child's chance of getting diabetes. When a mother has too much glucose in her bloodstream, glucose passes through the placenta and into the fetal blood supply. To compensate, the fetus uses its own developing pancreas to secrete more insulin and bring the blood sugar level down. But the extra work, and the glucose imbalance itself, can harm the not yet totally formed insulin-producing cells in the baby's pancreas, greatly increasing the child's chance of developing diabetes or hyperglycemia much later in life. Fortunately, obstetricians now know how to control a mother's sugar levels during pregnancy, and thereby can control the amounts of sugar the baby is receiving. Managing diabetic mothers during pregnancy is truly one of the success stories of modern obstetrics and clearly will have a major impact on the incidence of diabetes in future generations. For now, though, we are living with the results of the less-successful diabetes treatments of thirty or more years ago.

Pregnancy changes your metabolism, and even if you don't have diabetes normally, you can get a temporary form of diabetes (gestational diabetes) when pregnant. Even if you are not among the 2 to 3 percent of women who become diabetic during pregnancy, pregnancy can induce a periodic hyperglycemia that is not quite diabetes. At about the midpoint of pregnancy, your obstetrician will test for glucose intolerance by having you drink a high-glucose drink and measure how it affects your blood sugar levels one hour later. If you get an abnormal reading on this test, your doctor will generally ask you to take a more definitive test that requires pretest carbohydrate restrictions and blood tests every hour for three hours after drinking the high-glucose drink. Women with gestational diabetes generally return to normal after pregnancy, but they tend to have a higher risk of NIDDM later in life, which indicates that problems in handling sugar in pregnancy may suggest that something is not totally normal with the glucose management in their bloodstreams.

The current thinking suggests that women don't even need to have full-blown gestational diabetes to change the nutritional program in their child. Some authorities estimate that for every woman who has full-blown gestational diabetes there are many other women who have a greater than normal rise in blood sugar after eating, but not high enough to meet the criteria for diabetes. This greater-than-normal rise in blood sugar may have the same damaging effects as the blood sugar spikes in diabetes.

Ultimately, when we talk about obesity and diabetes, we are talking about the extremes of problems that are much more widespread: people being overweight and unhealthy. All the evidence suggests that these problems are not just promoted by diet and lifestyle factors among adults and children, but by the nutritional environment in the womb. If it is true that a fetus's nutritional program can be set in a far from optimal way by bad conditions in the womb, the opposite is also true. You can help your baby avoid these problems by a positive nutritional program through better eating. Right now, physicians clearly recognize the risks of full-blown diabetes during pregnancy and aggressively treat it. I hope that in the future they will recognize that obesity and other health problems are also big problems during pregnancy, and will treat them just as aggressively.

GOOD NUTRITION IN PREGNANCY
BEGINS BEFORE PREGNANCY

As with so many other activities in life, preparation is the key to giving your baby the best nutrition during pregnancy. The old saying "If we fail to prepare, we prepare to fail" may seem a trite adage, but it is abundantly true for every aspect of pregnancy. Although there is much conflicting advice on how to eat during pregnancy, all nutritionists agree that preparation for good nutrition during pregnancy begins before you become pregnant.

Two recent panels of high-level experts reached the same

very important conclusion, namely that the dietary rules for pregnancy are very much the same as those you should be following before you become pregnant. The US Department of Agriculture (USDA) has recently published a very readable and authoritative guide entitled *Nutrition and Your Health: Dietary Guidelines for Americans 2000, 5th Edition*. This up-to-date guide is available free on the web at www.usda.gov/cnpp. The complete 40-page booklet can also be obtained at $4.75 per copy by calling the Federal Consumer Information Center at (888) 878-3256 and asking for the booklet (item 147-G). I can strongly recommend this booklet to every woman, to every family contemplating pregnancy.

The Institute of Medicine, part of the National Academy of Sciences, has reached similar conclusions to the USDA and is in the process of producing multiple large volumes on dietary reference intakes that come to similar conclusions. Researchers are learning so much about nutrition so quickly that there is sometimes conflicting information out there, so it is reassuring to have clear, authoritative advice that is truly up to date as of the beginning of 2001.

The USDA guidelines summarize the key goals that they call the *ABC's* for both women and men. Women should follow these guidelines even before they think about pregnancy. We should *Aim* for fitness, *Build* a healthy base of nutrition, and *Choose* our foods sensibly. We will consider the *A* of the *ABC*—exercise needs—in chapter five, so let's look at the *B* and *C* here.

The first thing to do is to find out whether your nonpregnant weight is appropriate for your height. Determining whether you are overweight clearly requires you to know whether your weight is appropriate relative to your height. Healthy, tall individuals will generally be heavier than healthy, short people. To correct for differences in height you need to know your body mass index (or BMI). You can calculate you own BMI by dividing your weight in kilograms by your height in meters squared. To convert, divide your weight in pounds by 2.2 to get kilograms and multiply height in inches by 0.0254 to get meters. If you don't have a calculator in hand to do that math you can find BMI calculators on the Web by searching under BMI. A BMI greater than twenty-five suggests that you are overweight and that it would be wise to try to lose some weight before you become pregnant. A person with a BMI over thirty is considered to be obese. Obesity greatly increases your risk of ill health, especially from heart disease and diabetes. It is also not the best place to start a pregnancy. The best way to keep your BMI in the normal range is to balance the calories you eat and your level of physical activity. If you are putting on weight, the math you need to know to reverse that trend and reduce weight is simple: Take more exercise or eat fewer calories, or a combination of both.

EATING RGHT BEFORE YOU BECOME PREGNANT

Nutrition experts have developed a food guide pyramid to help you eat the right balance of the different foods you need to build a healthy body. No one food contains everything you need. To help you select food products with the right ingredients, the FDA requires that all foods have a nutrition facts label. Learn how to read that label. It tells you the various components in a product and the percentage of your daily needs the product fulfills. The number of servings in the container and the size of one serving are also listed. There is some confusion as to the use of the word *serving*. The USDA uses servings that are generally less liberal than you will find on the nutrition facts labels, so the nutrition facts label is yet another example of the many ways in which we are so often encouraged to overeat. As you develop your own diet, stick to one system, preferably the USDA dietary guidelines system that I will use here. Again, the important thing is to keep the right balance of foods from the pyramid. That way you will get all the nutrients you need. You can easily tell whether your servings are too large, or too many, for your body size and activity level. If the servings are too large, you will put on weight, often in places you do not want it. The key is balance. If you are a big person, you will need to increase the number of servings of foods from each level of the food pyramid equally. But you should keep the balance between different food groups the same.

The food pyramid is securely based on the grains group—especially foods that contain whole-grain. This group of foods includes good whole-grain bread, cereals, rice, and pasta. Depending on your height, you should be having six to eleven servings of foods from this group.

In some ways carbohydrates have gotten a bad rap. Carbohydrates have been the basis of the Western diet for many thousands of years, and provide us with sufficient energy to carry one through the day. We need to eat enough carbohydrates to keep us from scavenging our own proteins for energy. Extremely low-carbohydrate diets like the Atkins diet, so recently in fashion, are not the way to control your eating, either during or before pregnancy. Such diets promote exactly the sort of protein scavenging that takes protein building blocks away from the baby.

On the other hand, many people do eat too much carbohydrate, and specifically eat too much processed and refined carbohydrate. The starch in flour, rice, potatoes, and other carbo-hydrates is composed of chains of glucose molecules, and when the starch chain breaks down in the stomach, the glucose released quickly increases blood sugar levels. Your pancreas—and at the same time your baby's pancreas—releases a pulse of insulin to bring the blood sugar level back in line. But if you have gestational diabetes or even have a problem with hyperglycemia, blood glucose will stay high for longer than normal. This is not good for either you or your developing baby. Fats interfere with insulin activity, so a meat-and-fries diet can make hyperglycemia worse.

Eating whole-wheat breads, brown rice, and other foods high in fiber combats a sudden rise in blood-sugar following a meal and smooths out the blood sugar peaks and valleys. Fiber not only acts as filler to keep the volume of carbohydrate you eat down, it also slows the release of glucose into the bloodstream. Eating foods high in fiber has the added benefit of lowering your risk of colon cancer and of fighting constipation, which is often a problem in pregnancy. But our main focus here is in providing the best nutritional environment for your baby—eating six to eleven servings per day of whole-grain carbohydrates keeps the developing fetal pancreas from having to work too hard. A cau-tionary note: Don't try to go straight from eating only white bread to eating only 100 percent bran muffins or only brown rice. The high-fiber food will likely seem less palatable at first, so a gradual changeover is easier to take. Obviously, it would be best if you start to make that change before conceiving, but as they say, "better late than never."

From the solid base of the grain group, we move upward on the pyramid to a narrower level that is occupied by two groups:

the fruit group and the vegetable group. The recommendation is for three to five servings of the vegetable group. Again, the number of servings you need depends on your height—not your weight alone. Basing your intake on your weight alone can just compound the problem of feeding your body fat and becoming even fatter. As a good source of vitamins, minerals, and all-important fiber, vegetables have much to offer. Vitamins and minerals found in vegetables are essential for maintaining your own tissues before pregnancy. During pregnancy they are as essential to putting together tissue in your baby's growing body as thread is to sewing. Go for leafy, dark green vegetables first—they are an essential part of a good diet. Fresh vegetables are the best, but in a pinch canned or frozen vegetables will do.

You should eat two to four servings of the fruit group each day. While a great source of fiber, vitamins, and minerals, fruits have the added benefit of being sweet and snacklike. Many of the sugars in fruits have to go through a few extra steps to become glucose, so fruits don't cause as big a spike in blood sugar as refined sugar does. Fruit is a great thing to reach for instead of potato chips or candy bars. Again, fresh is absolutely the best, but unsweetened canned fruit counts too.

Ascending the pyramid, the third level also contains two groups: the milk group and the meat and beans group. The recommendation is two to three servings for each of these groups. Proteins are the basic building blocks of the body. They are the stuff that holds everything together and makes the machinery of life work, both for you and your baby. With so much new tissue being created, your baby will need a lot of protein. Too little protein will stunt your baby's growth and lead to a higher risk of disorders like diabetes. On the other hand, you don't need to have tremendous amounts of protein—extremely high protein diets have also been associated with poor health. Try to eat lean meat like chicken with the skin cut off, fish, and lean cuts of beef or pork. Legumes (dried beans or peas) are also a good source of protein. Variety is essential in choosing food for many reasons. Fish is an excellent source of many important nutrients. There has been recent concern that some tuna may contain mercury and that even farm-reared fish may contain toxins absorbed from the water in the hatcheries. It is probably wise to eat food from many different sources. However much you like salmon or herring from one particular area, fortunately you can get it from many parts of the world. Ring the changes.

Dairy foods (cheese, yogurt, milk, etc.) are excellent sources of protein as well as calcium. Much of the calcium that you eat during pregnancy will go into making strong bones in your baby during the second half of pregnancy. When eating high-calcium dairy foods, try to stick with low-fat types.

At the top level—the pointed peak of the pyramid—come fats, oils, and candies. The recommendation is that we consume representatives of this group sparingly. That can be hard to do since most of us like candy and fat gives much of the taste to food. However, much of the impaired quality of life that many people endure today is because they are overweight. We read daily that both the developing and developed world suffer from obesity. It is critical that you know the difference between the various forms of fat. High-fat dairy products such as cream, butter, and cheese and some oils such as coconut oil are full of the bad form of fat—saturated fat. It is very important to restrict your intake of these fats. Cholesterol is present in egg yolks and many dairy fats. Trans-fatty acids present in hard margarines and many fried foods are harmful because they raise blood cholesterol. The good type of fat is unsaturated fat, which we can obtain in fish and many vegetable oils. We need unsaturated fat for our cell membranes and especially to make the lining sheath of our nerves. This sheath is vital to normal nerve function, rather like the insulation around a wire in a cable. So unsaturated fats are critical for your baby when you become pregnant and it is wise to incorporate them into your dietary patterns early in life. There are some fundamental rules that will help you eat the healthiest type of fat diet: Use vegetable oils whenever you can; eat two to three servings of fish, nuts or lean meat daily. The amount of fat you can safely eat each day is related to your total calorie intake. The percent daily values on the nutrition facts labels on food products are based on a 2000-calorie daily intake. At that level of calorie intake, recommended daily saturated fat is less than 20 grams and total fat 65 grams. With this in mind it is an eye-opener to know that one croissant contains 6.6 grams of saturated fat while a bagel contains 0.1 grams. This staggering difference shows the critical importance of choosing the right food. Sugar and other sweeteners are also greatly overconsumed in Western society, but they too are not evil substances to be completely banished from the diet. Cut down on sugars where you can, and have fruit for a snack instead of chocolate or other candy. Artificial sweeteners generally seem to be acceptable in small amounts.

My advice is that you choose the most nutritious foods that you like, check out their calorie and nutrient values, and then go through the USDA dietary guidelines in order to make yourself a daily diet sheet, allocating different foods to each meal. You can then add up the number of servings you are getting in each of the food groups as well as your daily calorie intake. There is no one correct diet, regardless of what the pundits say about their own special formula. Your own best diet will inevitably reflect many factors in your life: your culture, cost and availability, allergies and many other personal features. However, do ensure that the diet achieves the balance that exists in the pyramid. Choose the best quality foods that you can afford in each group. There are virtually always healthy options open to you. If you cannot eat—or choose not to eat—dairy products, you can get your calcium from foods that have been fortified with calcium such as juices or cereals. Once you have chosen the foods you like, then you need to read the nutrition facts labels. When reading the label ask yourself if your choices are providing good nutrient value for the calories you will be taking in. Remember, the golden rule is that once your diet exceeds the number of calories you burn each day, you will inevitably put on weight. Next, look to see how much saturated fat the product contains. It is best to get used to looking at the absolute amounts as well as the daily value, since clearly the absolute amount that constitutes your own personal daily need differs according to your size. Also you should remember that serving sizes may differ from the serving sizes recommended by the USDA.

SPECIAL COMPONENTS OF A HEALTHY DIET

Vitamins and minerals are substances essential for life, but our bodies either can't make them at all, can't make as much of them as we need to be healthy, or cannot store them. We have to get them through our diet. Some vitamins are more common in some kinds of foods. That's one reason I recommend a varied diet: eating many different kinds of foods is more likely to give you a full complement of vitamins and minerals. Always read the nutrition facts labels to get an idea of how much of the percentage daily values of each vitamin and mineral you are consuming in a serving of the product. Orange vegetables like carrots and pumpkin are good sources of vitamin A. Broccoli, leafy greens, and citrus fruits are good sources of vitamin C. Folic acid is present in beans, peas, and peanuts, orange juice and leafy vegetables. It is important to note how often vegetables crop up as good food. Some vegetables like spinach also contain important minerals such as potassium that our cells need to conduct their active functions. The general recommendation is that women should not take vitamin supplements—unless their doctor recommends them. Instead, women should get their vitamins and minerals through their diet. Possible exceptions are for iron and folate. Too much of some vitamins, such as vitamin A, can actually cause birth defects. All vitamins and minerals are important, but some are particularly important in pregnancy:

Calcium is the main building block of bones and teeth, and is also critical for the proper function of your own and your baby's nerve cells and muscles. You need 1200 to 1500 mg per day. Calcium is primarily found in dairy products like milk and cheese, but is also concentrated in broccoli, kale, legumes, and tofu. Calcium absorption from food increases during pregnancy. Although some doctors recommend calcium supplements to make sure that women are getting enough, the Institute of Medicine's recommendations on nutrition in pregnancy state quite clearly that if you are on an adequate diet, calcium intake does not normally need to be increased during pregnancy. If you are short of calcium, the best way is not to take extra calcium or vitamin D as a supplement. The committee recommends eating more milk, cheese, or yogurt, which are the forms of food that provide other important nutrients in addition to calcium such as protein, and other minerals and vitamins. Indiscrimately loading up with supplements can cause its own troubles. Some minerals interact with each other. For example, calcium can interfere with the absorption of iron.

Your iron needs during pregnancy are 30 to 60 milligrams a day in order to allow your blood supply to increase and help your baby build her own blood supply from scratch. Iron plays a central role in providing oxygen to your baby. It is ike a jewel at the center of the hemoglobin molecule, latching onto oxygen and carrying it through the body. Without iron, blood cannot carry oxygen, so all new blood must have sufficient iron available when it is being made. Lack of iron in the diet causes anemia—in both mother and baby. Iron is found in red meat and especially in liver, and is also found in eggs and dried beans. If you are short of iron, your physician can prescribe iron supplements during pregnancy. These iron pills are best absorbed on an empty stomach, but taking them with a little food is fine if the pills tend to make you nauseous. Warning: Keep iron pills out of the reach of children. Iron may seem like an innocuous substance, but an overdose is extremely toxic and can be deadly.

Like iron, zinc is a trace element that is crucial in pregnancy. You need about fifteen milligrams a day. It is important for tissue growth and the reproduction of genes in DNA. Zinc deficiency during pregnancy can cause low birthweight or improper brain development. Zinc is found in whole grains, nuts, dried beans, meat, and eggs.

•   •   •

Chromium, another trace element, doesn't get a lot of attention, but it helps insulin work to keep your baby's blood sugar at the right level. Chromium deficiencies can make blood-sugar levels higher than normal after meals. Chromium is found in whole grains, meats, and brewer's yeast. These sources can easily supply the 50 to 200 micrograms that you need each day.

Since folic acid (the synthetic version of folate found in foods) is vital for the synthesis of DNA, it is not surprising that every cell in your baby's body, and the placenta, needs an adequate supply. Having adequate folic acid before pregnancy (400 micrograms every day as part of a healty diet) has been shown to lower the risk of neural-tube defects by up to 70 percent. It is necessary to take folic acid before pregnancy because neural-tube defects occur in the first 28 days of prenancy, often before most couples know the prgnancy has occurred (the 28th day of the pregnancy is two weeks after the first missed period or 6 weeks after the last period). Physicians recommend that women who have had a child with a neural-tube defect (who are 10 to 20 times more likely than the general population to have another child with a neural tube defect), or are at higher risk for a neural-tube defect start taking folic acid (4000 micrograms per day, 4mg) for four months before conception (if the conception is planned), and then during the first three months of pregnancy. Because folic acid is essential for the formation of red cells in the mother and the fetus it is included in prenatal supplements if they need to be continued throughout the pregnancy. For women who are not at higher risk for a child with a neural-tube defect the recommendation is to consume 400 micrograms (0.4 milligrams) of folic acid daily as part of a healthy diet (this is the recommendation of the Food and Nutrition Board of IOM from 1998, and is similar to the USPHS recommendation from 1992). In the United States you are probably getting a fair amount of folic acid in your daily diet because the Food and Drug Administration requires grain mills to enrich flour with folate. Folate is also present in fresh fruit and green vegetables, another argument for making sure you get these foods. Supplements may be also be necessary in adolescent pregnancy, if you are carrying twins or for those who still must smoke regularly despite the known adverse consequences to the baby.

Vitamin A helps build key components of your baby's skin, eyes, and other tissue. But beware: too much vitamin A is harmful to the fetus and can cause birth defects. This is a classic example of a case in which you can get too much of a good thing. Since vitamin A comes in several forms it is usually measured in units. The recommended daily intake is around 800 international units (I.U.)

Vitamin B_6 will help your baby create the new tissue he needs to grow, especially the billions of nerves that make up his brain. B_6 is found in a wide variety of foods, such as eggs, whole grains, lean meat, oatmeal, nuts, dried beans and peas, and bananas. The amount of vitamin B_6 your body needs increases from around 1.6 mg before pregnancy to 2.2 mg a day at the end of pregnancy.

Vitamin B_{12} plays a central role in the production of new DNA your baby needs as her cells multiply. Good sources of vitamin B_{12} are meat, fish, eggs, and cheese. Your needs will increase from 2.0 to 2.2 micrograms a day during pregnancy.

Vitamin C is crucial for cell repair after injury and for the development of new tissues. However, you don't need to take massive doses to get enough. In fact, the recommended daily intake during pregnancy is 70 mg, only slightly above the normal 50 mg per day recommendation before pregnancy. Fruits (especially citrus), and dark green vegetables are good sources of vitamin C.

DIETARY SUPPLEMENTS

We are a nation addicted to the quick fix. Every large grocery store has its section of shelves stacked with pill boxes of all shapes and sizes. We are assailed with seductive marketing information for vitamins and mineral supplements. The golden rule is that you can fashion a good diet that does not need these added ingredients. But, you may say, surely it is playing safe to take more? As we have seen with vitamin A, the philosophy of "more must be better" can be a very dangerous way to go both before and during pregnancy. Every hospital internist has their favorite story of a patient with kidney stones who thought that taking yet more and more calcium and vitamin D could only be beneficial only for the calcium to precipitate out in their urine and form a stone. Some people do need supplements, especially as they pass middle age, but that's another story. Certainly, as already mentioned, pregnant women need folic acid, especially before pregnancy and in the early pregnancy stages. Some women will need iron if their blood hemoglobin levels are low in pregnancy. Apart from folic acid and iron the consensus now is that supplements are not necessary in pregnancy if a woman has a properly balanced diet. The human race reproduced successfully before the days of superstore multivitamins.

I might add that the picture of the pyramid should include a lake. That is because most people do not drink enough fluids. When you are pregnant you should be drinking a lot of fluids every day, since you are clearing out your own body's waste as well as the baby's. The general guideline is that you should drink about eight cups of fluid (mostly water) a day. Somehow this recommendation gets altered in most books to say eight "glasses" of fluid per day, and since most tall glasses are about twelve ounces instead of the eight in a cup, many women find fulfilling this requirement difficult. It should not be, if you think in terms of cups and count all the water and juice you drink throughout the day. Drink more if the weather is hot or you are exercising.

VEGETARIAN MOMS

Vegetarians face special challenges in fulfilling their nutritional requirements during pregnancy. Most vegetarians know well what they have to do to get their normal daily requirements, and in fact probably tend to have better diets than nonvegetarians because they have to think carefully about what they eat. However, if you are a vegetarian, you will have to be even more careful about what you eat to fulfill your developing baby's nutritional requirements as well as your own. There are, of course, many different kinds of vegetarians: those who eat no meat but do eat fish, those who don't eat dairy products but do eat eggs (or vice versa), and vegans, who don't eat any animal products at all, are only a few examples. Each type will have to tailor their diets so that they meet minimum requirements while meeting their chosen dietary proscriptions.

Nonvegetarians may be surprised that getting enough protein

is not the biggest problem for most vegetarians. Protein consumption does need to increase during pregnancy, but vegetarians tend to know what kinds of foods they need to eat to meet that requirement. The bigger challenges for vegetarians are getting enough calcium, iron and, vitamin B_{12}. This is especially true if you don't consume milk and/or eggs. Vitamin B_{12}, for instance, is only found in animal products, so vegans will not get enough B_{12} without taking supplements. The answer is to check labels carefully and add up listed nutrients to make sure you are getting the recommended amounts for pregnancy. If not, you may have to consider modifying your diet or taking supplements (after consulting your doctor) during pregnancy.

Vegetarians are not the only group of individuals who need to take special advice about how their individual lifestyle may affect their dietary needs. The active form of vitamin D is produced in the skin from precursors by the action of sunlight. Vitamin D shortages can occur in pregnant women in communities where women are covered all day and get no exposure to sunlight.

ESTABLISHING THE GOOD EATING HABITS THAT YOU NEED DURING PREGNANCY

Eating right doesn't have to be the enemy of eating well. In other words, healthy eating doesn't mean you can't enjoy eating. And it doesn't have to be a huge burden. You don't have to eat the huge portions of food or drink the buckets of water that many pregnancy books seem to suggest. Rather, the principles of healthy eating both before and during pregnancy are fairly straightforward. Eating right means:

- Knowing what constitutes healthy eating.

- Planning so that you aren't just grabbing what's available.

- Procuring the right foods so that you have the right things to eat when you need them.

The tough part is making a commitment to those principles and putting theory into practice in everyday life. *Habit* seems like a lightweight word. It seems to imply a casual allegiance to some action or thought. But habits are very hard to break. Eating habits are some of the most difficult to change. In large part this is because food has meaning. We don't eat just for food's nutritive value, or we could all just consume glucose, fiber, and pills for vitamins and minerals. We eat to have a certain sensation, to feel certain emotions. We like coffee with the morning paper, popcorn with a movie, a glass of wine with a sunset. Eating a favorite food is emotionally rich and satisfying. It carries memories of past moments when we savored that food as well as the current sensations in the mouth and stomach. We want comfort food because we so much want to be comforted.

If the foods we most often choose are not quite the healthiest choices, it can be uncomfortable to change, to forgo the snack for the less comforting but higher principle of healthy eating. It can be particularly difficult to make that change if you are already so uncomfortable, if you are feeling physically and mentally knocked about. When you become pregnant and so much of you feels different, you especially need the reassurance of comforting food. So for many women, changing eating habits is more easily accomplished before pregnancy.

On the other hand, many women find that becoming pregnant is the spur that they need to get serious about their health habits. Despite the discomfort of pregnancy, they now have direct responsibility for the well-being of someone other than themselves. They take strength from that knowledge and are motivated to do all they can to eat the right foods.

The guidelines for healthy eating during pregnancy are, for the most part, simply the usual rules for healthy eating in all adults, with adjustments for both the particular needs of a mother who is pregnant and a growing fetus:

- Start eating right before pregnancy.

- Keep your own body healthy—food goes to the baby through you after minor processing by your own body.

- Eat balanced meals, with a little more protein than usual.

- Adopt a strategy of eating small portions often, and plan so that healthy "minimeals" are there when you need them.

- Don't try and keep your prepregnancy weight.

- Don't let yourself gain too much weight.

- In the second and third trimester, you will probably need about 300 more calories per day than before you were pregnant.

- Eat food high in iron and vitamins, and as a second line of defense if your caregiver recommends them, you can take pregnancy vitamin supplements.

PREPARING FOR PREGNANCY

Whether you know it or not, your preparation for pregnancy was begun long ago while you yourself were in your mother's womb, shaped by the preparations she made for pregnancy and the messages that she sent you across the placenta. The womb environment that your baby will experience was further nudged in one direction or another by how you ate in your teen years, and the kind of exercise you got. At each stage of your life, your body has made decisions about how it will grow, based on the information you have supplied through the kind of life you've led.

In the next chapters, I will discuss about three fundamental aspects of the womb environment that are critical for programming lifelong health for your child: moderate exercise, avoiding toxins, and alleviating stress. Each of these is important during pregnancy, but they are also equally important to address before pregnancy. When conception occurs and pregnancy begins, the body is already communicating chemical messages based on its current state of health. You have no control over how you developed in your mother's womb, and you can't change how you behaved as a teen, but you can get your body ready now.

As we have seen, in the best of all possible worlds, you should begin preparing for pregnancy well in advance of conception. The important thing is to begin as early as possible, whether you are pregnant now, just thinking about being pregnant in the future, or planning on getting pregnant soon. In fact, nutrition before you get pregnant is just as important as nutrition during pregnancy. Nutrition in the first few weeks after conception—when the embryo is forming all of the most important tissues and organs and you may not even know that you are pregnant—is critical for a healthy pregnancy and a healthy child. What you eat *before* pregnancy begins sets the physical foundation upon which you will build your child. Conceiving is like planting a seed in a garden. It's best to plant in the right season, when the soil is moist and rich with the right nutrients. The condition of the soil before planting has as much to do with growing healthy plants as how well you take care of them after they start to grow. If the soil is naturally poor quality and not well prepared with soil conditioners, plants won't grow well, despite faithful watering and fertilizing. In pregnancy, the conditions in the blood-rich tissues of the womb bed before fertilization will be vitally important to the future growth of the embryo. For instance, how your body is processing and metabolizing food before pregnancy will determine the makeup of the womb lining to which the newly conceived embryo will attach itself. Prior to the development of the placenta—a critical developmental phase for the embryo—the embryo will feed on the glucose, amino acids, and micronutrients already in the womb lining and its secretions.

NUTRITION DURING PREGNANCY

You don't have to eat all your food within a standard breakfast-lunch-dinner format. In fact, in many ways it is best if you spread the food out by creating four, five, or six smaller meals every day. Snacking between meals is not only okay, it is encouraged as long as you stay within the daily totals for healthy eating. Spreading the daily ration out is particularly helpful in early pregnancy because you keep food in your stomach, keeping down stomach acid that can make you feel sicker. If you are feeling nauseous during some times of the day you will probably be eating sporadically, taking advantage of the times when you actually feel hungry. Keep a little piece of bread or cracker near your bed to get something in your stomach right away in the morning. Keep a healthy snack such as fruit in your purse. Keep the saltines nearby at all times. The important thing is to plan snacks and prepare them beforehand so that when you can eat you do so in a healthy way, within the plan.

During the second and third trimesters, spreading the daily meals out also makes sense because your stomach is getting squeezed as baby pushes out. You can't eat as much at one time, so you are forced to eat smaller meals more often.

NUTRITION BY TRIMESTER

FIRST TRIMESTER

In the first trimester, your biggest jobs are

- Building a healthy placenta.
- Dealing with the effects of morning sickness.
- Making sure all essential nutrients and sufficient fluids are present in your body.

During the first trimester your microscopic embryo, having attached itself to the wall of the womb, begins to grow the placenta that will provide support and nourishment for the rest of the pregnancy. The growing placenta requires a good supply of protein. And even though a very low calorie diet, as I mentioned before, can spur the growth of a larger placenta to make up for the shortage, it is not a good idea to skimp or diet at this point. For one thing, if you do not have enough calories available from glucose, your body will scavenge proteins to make energy, thus stealing protein from the growing placenta and embryo.

On the other hand, you don't have a license to eat anything or as much as you want. I knew one woman who had spent all her life worrying about her weight and suffering under strict diets. So when she became pregnant she heard a clarion call to eat with impunity, since everyone is expected to put on fat during pregnancy. I had to let her know that this was the wrong approach, that she still should not binge. However, she was happy to hear that she should definitely not be dieting like she had been before. I told her that she should eat sensibly and let her body find her natural weight.

If you are eating well before pregnancy, there is no need to actively increase your calorie intake at the very beginning of pregnancy. You should only gain a pound a month in the first trimester. Gaining weight too quickly can lead to high blood pressure, which in turn can lead to preeclampsia, which can become a life-threatening condition. Overeating can also make existing blood sugar problems worse, with consequences for the baby. High blood sugar in mom leads to high blood sugar in baby. Fetuses exposed to continually high levels of blood sugar may grow very large. The research that I have mentioned on the revolutionary idea of fetal programming of adult disease has been mostly conducted by studying adult health in individuals who were underweight babies, since that is by far the more common problem. However, there are also indications that babies at the high end of the weight scale are not programmed for optimal health either. For example, there is a correlation between high birth weight in females and breast cancer in later life.

A healthy womb environment demands not only growth of an efficient placenta in the first trimester, but also development of a good maternal blood supply carrying a full complement of nutrients to the baby. That means that you should take plenty of iron-containing foods for the growing blood supply, and eat balanced meals. A shortage of vitamins or minerals at this point can have a negative effect on tissues during critical points of development.

Folic acid is a clear example. It is very important early in the first trimester because this is when the spinal cord is being created. Iron is crucial for the expansion of the blood supply.

Your biggest challenge at this time is dealing with the effects of morning sickness. Morning sickness is not necessarily restricted to the morning (some people experience it all day) and can cause such severe vomiting that women require hospitalization and intravenous hydration. Eating may seem like the last thing you want to do when you feel so terrible, but keeping something in your stomach can actually help. As recommended earlier, keep a small snack next to your bed to eat first thing in the morning and then eat small meals throughout the day. You will probably find it helps to avoid strong odors like cigarette smoke, onions, and garlic and to find pleasant odors like fresh lemon.

If morning sickness causes vomiting, you need to replace what your body has discarded. Vomit drains your body not only of fluids but also essential salts like potassium. Drink extra water if you are vomiting a lot. Salts can be replaced with the food in a well-rounded diet. Sports drinks are spiked with a variety of salts to replace those that are lost through sweat, but they work equally well for salts lost through vomiting. I don't recommend making these drinks your main source of fluid, however, because they also contain a lot of glucose.

The healthy partner to good diet is exercise. In chapter 5, we will consider why exercise is good and will discuss the kinds of exercise that create a healthy womb environment during pregnancy. Even though you are feeling tired and sick during the first trimester, exercise can help the body adapt to your changing metabolism and ease morning sickness.

SECOND TRIMESTER

In the second trimester, morning sickness may begin to fade and you will start to feel more energetic. Finally! Your abdomen and the baby inside really begin to grow at this point. You should be eating about 300 calories more per day, but not much more. Again, your daily weight will tell you whether you are eating the right overall amount.

Protein continues to be essential in the second trimester, more for building your baby's body than for building the placenta. Your blood supply is not expanding as quickly as it did during the first trimester, but the baby's blood supply is growing along with her growing body, which means that iron is still crucial. Calcium is especially important during the second and third trimesters, as the cartilage framework is transformed into a true, bony skeleton. Bone is a living tissue that needs to be nurtured and fed just like other organs, and getting enough milk, cheese, and yogurt will make it possible for your baby to grow tall and strong. If you don't consume dairy products because you are lactose intolerant or for other reasons, you will have to work harder to make sure you get enough calcium, perhaps through supplements. Eating high-calcium foods like legumes, broccoli, molasses, eggs, and others listed above is the first option to try, however.

Even if you don't have gestational diabetes, you may have some episodes of hyperglycemia that are exacerbated by pregnancy. If you tend to suddenly become ravenously hungry and shaky at moments during the day, then feel wiped out and drained of energy after eating, you may have unhealthy blood sugar fluctuations. Don't make your pancreas and your baby's pancreas work harder than they should. Don't expose yourself or baby to higher than necessary blood sugar levels. Keep trying to eat small snacks more often. Eat carbohydrate products made with whole wheat, bran, or other fiber sources. Avoid fatty or oily foods because fats interfere with insulin activity. Getting starved and then eating a large helping of food with a lot of fats and carbohydrates—like a big slice of chocolate cake or a bowl of olive-oil–soaked pasta—makes blood sugar spike and is just about the worst thing you can do.

THIRD TRIMESTER

During the third trimester, your baby becomes much more physically demanding. He is growing rapidly and needs more nutrients daily. As he grows ever bigger, your stomach—perched atop the uterus—grows ever smaller as it is squeezed higher and higher in your abdomen. Eating small meals often is still the rule, but now it is required more by the physical size of the stomach than by morning sickness (although some women still feel nauseous into the third trimester). Eating healthy snacks between meals also remains a good idea for women who are borderline hyperglycemic after eating.

Building those bones is still important, so getting enough calcium continues to be a focus in your diet. Even if you are lactose intolerant, you will find that you may be able to take dairy products a little more easily than usual in the third trimester. How women become more lactose tolerant during pregnancy continues to be a mystery, but we can make a good guess about the reason why: calcium is so important to a growing fetus that evolution has crafted a response that allows women to get that calcium more easily. Another evolutionary adaptation may be pica, a craving or desire to eat clay, chalk, or dirt during pregnancy. Pica may be another attempt by the body to get sufficient minerals that may be missing from the diet.

You should continue to gain about a pound a week, but if you have been gaining more than that during the second trimester, you shouldn't try to make up for it by trying to lose weight now. You and your doctor may decide you need to cut back a little in total caloric consumption, but never try to lose weight during pregnancy. You should always be getting bigger—it's a natural part of the program.

Even when you swear you cannot possibly expand any more you will gain another couple inches in girth. This presents challenges for exercise, but you should continue to try to meet your goals. After all, you will benefit if you are ready for the exercise involved in labor. You need to keep in training for it. We can look at what the experts say about exercise throughout pregnancy, including the third trimester, in chapter 5. You may also find that you have less energy for shopping for food and cooking it, but this is not the time to let your nutritional goals slide. This is when planning, writing down meal plans, and keeping cold snacks handy can pay off.

REMEMBER, PERFECTION IS NOT POSSIBLE

Just as you shouldn't aspire to creating a "perfect" womb environment, you also shouldn't try to make every bite part of the perfect diet. Work on improving your diet, but don't let yourself get wrapped up in guilt over every lapse. Like all diets, you need to allow yourself some latitude, some rewards. You won't be barred from heaven if you occasionally enjoy a sinfully delicious dessert. You won't be stunting your baby's growth if you are occasionally forced to skip a meal or you don't have a balanced diet for one day. We are a very resilient species. There is some forgiveness in the system. The body has a way of storing up needed nutrients and energy to carry it through brief periods of shortage. If humans were so fragile that we couldn't stand to get hungry for a few hours once in a while, our species wouldn't have made it this far. The important thing is to learn what's good for your baby and then work to incorporate the principles of healthy eating into your life.

One final reminder about the importance of a good, well-balanced prepregnancy diet. There were many outstanding researchers who study nutrition in pregnancy on the Food and Nutrition Board of the Institute of Medicine and the USDA group whose recommendations I have summarized here. These experts repeatedly stated that the way to get the appropriate amounts of various nutrients is to eat a good balanced diet replete with fruit and fresh vegetables and a balance of protein and complex carbohydrate. Food supplementation, with the exception of folic acid, which has been clearly shown to decrease the risk of neural tube defects, in their opinion, was only necessary when there was a deficit prior to pregnancy. It needs to be said time and time again that giving the gift of good health to our children requires fathers and mothers to plan, look after their own bodies before pregnancy, and seek out experienced care givers so that together they can monitor progress throughout pregnancy with the powerful tools that modern biomedical science has put at our disposal.

4

Stress in the Womb:
Smoothing the Nine-Month Ride

A few years ago I was working with a woman—I'll call her M—who had a highly demanding position as a new faculty member at another university. Despite the calm, scholarly appearance of most university campuses, the pressures on new faculty to simultaneously do fine research, publish important papers, develop a curriculum, and teach daily classes are tremendous. These days, if new faculty don't contribute to the glory of their university, administrations can be as ruthless as any corporation in cutting them loose. In addition, M had the "curse" of connecting well with undergraduates, and so was pressured to take on the job of dean of student affairs. When she became pregnant, she began to worry that the amount of stress she was under might harm her child, but she felt unable to cut back on any of her responsibilities. I appreciated M's worries and worked with her to lessen her stress.

I wish I could tell you that I simply provided her with a powerful stress-busting method and she began to feel as relaxed as a cat lying in the sun. As is true in most cases, it wasn't that easy. M tried simple physical exercises. She tried focusing on her breathing. She tried drinking warm milk. None of these techniques was effective. In some ways they added to her stress by making her feel that she had even more tasks to fit into her schedule. The suggestion that finally worked for M was to take a few minutes a few times a day to imagine her daughter (she knew by then that she would have a girl) at a specific age in the future—six, fifteen, twenty-four—and think about what she might be like. This not only took M's mind off her huge responsibilities and gave her a sense of calm, but it let her put events into perspective. The difficulties she was dealing with receded in importance, and even a year after her daughter's birth, M reported that she was better able to deal with stress than she had ever been before. This technique won't necessarily work for you, but you need to find one that does.

In modern life, there is no escaping stress. If you are like M, your life probably seems rigged so that your schedule is always too full, and existing commitments can only be knocked off your to-do list by a more pressing commitment and more guilt. Even before pregnancy, you probably found that work, housework, paying bills, odd jobs, and social obligations more than filled your limited time. Once you are pregnant, it becomes even harder to cope. And yet the science of prenatal programming reveals that it is very important to your child's happiness that you not become overly stressed. Even though you may find it hard to slow down during your pregnancy, you may want to stop and consider how stress can affect your developing child:

- Excessive stress can put your pregnancy at risk.

- High maternal stress levels can have a lifelong effect on how your child's brain and body develop.

- High or low maternal stress can affect your child's temperament.

- High maternal stress can make your child overreact to stress in the future.

- High maternal stress can make your child more susceptible to depression and other mental disorders.

What we now know is that a highly stressful pregnancy will influence the environment in which the fetus develops, and will mold stress circuits in his brain and body. These altered stress circuits may program your child's brain so that he is less able to keep a lid on roiling emotions and more likely to let anger and frustration boil over when difficulties arise.

Chronic stress can also affect your own health and your ability to get pregnant in the first place. High stress not only interferes with ovulation, but it can also curb sexual desire and make successful fertilization even less likely. People under stress also tend not to eat very well. They tend to eat very little or too much, and rarely do they eat a balanced diet. Chronic stress also is dangerous for the cells in your body, and can be toxic to brain cells.

How should you feel about this knowledge? You might simply view this scientific finding as another source of blame, anxiety, and stress. I firmly believe that this does not have to be so. For one thing, only very high, prolonged stress levels are harmful. I am convinced, and have convinced many expectant mothers I have talked to, that knowledge about the effects of stress during pregnancy can be a source of empowerment. Like M, if you are planning ahead and are willing to chart a course with built-in stress relief, you and your partner will be able to build a foundation of health and happiness for *both* you and your child.

THE SCIENCE OF STRESS

To understand how stress affects the fetus, it's useful to know a little bit about how the brain and body react to stress. The concept of the pervasiveness of stress is expressed succinctly by pioneer stress researcher Hans Selye, who said, "Stress is life and life is stress." What he meant was that life is a process, a series of changes and challenges. We try to keep everything in balance— maintain our body temperature, ensure that we get enough of the right kind of food, stay on schedule, and keep ourselves happy. But whenever things are changing, as they always are, we have to react. We try to bring our world back into balance or equilibrium. Selye was saying that life cannot exist without change. Change and stress are part of being alive—and part of feeling alive. When we meet challenges successfully we feel happy and fulfilled. While the idea of living in a quiet room where everything is always the same sounds appealing during our most stressed-out moments, living there permanently would be unimaginably boring.

We need at least a few challenges to feel truly alive. Where these challenges, which we call stressors, are manageable, they make us feel good. On the other hand, when they feel overwhelming, we call it *distress*. Distress is the feeling of being threatened by stress, what we feel when we doubt our ability to return to a state of equilibrium.

To cope with threatening levels of stress, our bodies have developed a stress response, a revving up of the key body systems to give us a boost in coping with or escaping from the stressor. When faced with a significant stress, the stress response ensures that the body gets a boost of energy which heightens our degree of alertness.

For example: imagine you are coming out of a grocery store

into a busy parking lot leading a child with one hand and pushing a shopping cart with another. This is stressful enough, but let us further imagine that when you are standing between your car and another, you drop the child's hand for a moment to get your keys. At that exact instant she decides to dash across the roadway to pet a dog. This is what happens: Your eyes take in the danger and your brain relates the situation to your past experiences of busy streets. The nerve circuits that respond to fear shoot off like fireworks. These circuits do many things, one of the earliest of which is to send nerve signals down the spinal cord to the adrenal glands telling them to release adrenaline. Although adrenaline gets the most attention as the adrenal glands' response to stress, there is also another way the adrenals get into the act. At the same time as impulses are passing along the nerves to the adrenals, the brain alerts the hypothalamus—the part of the brain that governs the body's basic drives, like hunger, sleep, and sex. The hypothalamus then sends out chemical signals to a gland—the pituitary—that sits just under it. The pituitary then releases a different hormonal signal, which travels through the bloodstream until it finds the adrenal glands sitting atop the kidneys and instructs the adrenals to release the less well noted stress hormone cortisol, which is responsible for many of the midterm and long-lasting changes produced by a big scare. Cortisol belongs to the glucocorticoid family of steroid hormones, which performs a large number of jobs in the body.

When your child runs into the street, nerves from the brain and adrenaline whip the heart into a fast, pounding rhythm. The rapidly beating heart rushes blood full of adrenaline and cortisol quickly through the body. Fat, an excellent and concentrated source of fuel, is mobilized for extra energy. The liver dumps stores of glucose, another energy source, into the bloodstream.

Your breathing quickens. The increase in energy molecules and oxygen in the blood provide muscles with the raw materials for a burst of strength.

With a speed and strength provided by these physiological boosters, you push the loaded shopping cart aside, leap after your child, grab her by the collar, and lift her straight up and back off the roadway. Your body's reaction is sometimes called the "fight or flight" mechanism because your body would react the same way if you had been confronted by a predator in the wild or a mugger in the city.

The activity of the brain-pituitary-adrenal system determines how much cortisol we have in our bloodstream at any time. It is important that our bodies are not exposed to too much or too little cortisol over an extended period of time. Too little cortisol over a long period results in muscular weakness and an inability to mobilize energy in times of stress. But too much cortisol over an extended period leads to bone weakness, ulceration of the intestines, nerve damage, and greater susceptibility to disease.

Just as we need some cortisol at all times during postnatal life, the fetus needs cortisol during prenatal development. In the developing baby, cortisol plays a key role in the maturation of the lungs, liver, kidneys, immune system, and many other body systems that are crucial for life outside the womb. In the womb, babies provide cortisol for this job from their own developing adrenal glands.

Not all stress is bad for the fetus. When your cortisol levels are in the normal range, the placenta effectively blocks the passage of most of the hormone from you to your child. The problems come when the stress is frequently at a high level over days or weeks. How much stress is too much? If you feel good about the stress you are under, if you feel you thrive on stress, then you probably don't have a problem. I remember a pregnant woman, Patricia Murphy, who was a maternal fetal medicine specialist taking care of other women with very high risk pregnancies. She worked right up until a couple hours before she went into strong labor. She was a very levelheaded person who could work very hard without feeling overly stressed. In fact, she was the kind of person who felt stressed only when she wasn't doing her best. Not everyone is like that.

Stress that is damaging is uncomfortable stress—invoking feelings of being trapped, anxious, and eager to escape. In these situations, cortisol can leak through the placenta and affect the developing fetus. The fetus can also be exposed to too much of your cortisol if the placenta has not developed normally early in the pregnancy, perhaps due to insufficient nutrition.

Whatever the cause, when the baby is exposed to more than normal amounts of your cortisol, it can have permanent effects. Studies with pregnant rats have shown that fetal exposure to excessive glucocorticoids leads to problems. One of the most dramatic effects is on the brain—more specifically, on the way the brain circuits are put together, and how we react to stress postnatally.

THE BRAIN UNDER STRESS

Each of us has a different reaction to stress. Some people are overwhelmed by the slightest stress or alteration of their normal pattern of doing things. These people may have a "short fuse," or have general anxiety and always fear the worst. Others appear calm and collected under almost any circumstances. Stress certainly appears optional to them, or at least controllable to a level that they can easily tolerate.

Move highly stressed mothers → work →
more highly stressed children → more stressors (daycare,
strangers) + cortisol

90   THE PRENATAL PRESCRIPTION

What accounts for these differences? Clearly, some of the difference is due to genetic makeup. The sum total of all the good and bad experiences you have experienced throughout your life since you were born also goes toward determining your stress reactions. But several groups of researchers in a variety of laboratories throughout the world have shown that a fetus's exposure to excessively high stress hormones will permanently alter the activity of critical components of the stress machinery later in life.

In 1990, Dr. David Phillips and his colleagues tracked down over one thousand men who had been born in Hertfordshire in England between 1920 and 1930. Although these men were in their sixties and seventies when the researchers located them, the researchers had records of their subjects' birth weights. The researchers measured the amount of the stress hormone cortisol in their subjects' blood and, in support of the concept of prenatal programming, found that plasma cortisol was highest in those men who had been the smallest babies at birth. This meant that the babies who had been most highly stressed while in the womb—through either lack of food or high stress in their mothers—had not only grown more slowly and less than they should, but had also had their stress response permanently changed. The highly stressed environment of the womb had given them higher stress levels for the rest of their lives.

How high prenatal stress levels might permanently change someone's stress level was clearly shown in an experiment done with pregnant rats. Scientists have long known that when rats are restrained, they become very anxious. In one experiment, rats were regularly restrained during pregnancy, and then their offspring's stress reaction was tested after they had become adults. The offspring of the mothers who had been stressed during pregnancy had a much more vigorous reaction to being restrained and had a higher blood glucocorticoid concentration than the restrained offspring of rats who were not stressed during pregnancy. To confirm that glucocorticoids produced by the pregnant mothers were responsible for this change, the experiment was repeated, except that this time the adrenal glands were removed from the pregnant rats. After the glands were removed, restraint produced anxiety but not higher glucocorticoid levels. The offspring of these rats (which were not exposed to higher levels of stress hormone) did not have a more vigorous stress response when they grew up.

How can a big increase in the level of glucocorticoid in your blood permanently alter fetal development? The answer to the puzzle lies in the fact that cortisol is not only a part of the stress response, it is also part of what is called a negative feedback system that keeps the stress response from spinning out of control. Negative feedback works like a thermostat—if it gets cold, the thermostat turns on the heat, but the temperature will only rise to a preset limit before the heater shuts off. During a stressful situation, cortisol rises; but at a certain level, cortisol circulating in the blood tells receptors in the brain to turn off cortisol production. What researchers have found is that high cortisol levels in the fetus reset the cortisol "thermostat" so that it is less sensitive and therefore allows cortisol to rise to much higher, more harmful levels during stressful situations later in life.

To get a better idea of how this works, imagine that you have just moved into an apartment in which the thermostat for the heating system is placed on a very poorly insulated outside wall. As a result, when a cold snap sets in, the thermostat senses the outside temperature rather than the temperature in the room and switches on the heating full blast. Although the temperature in the apartment rises well into the eighties, the thermostat thinks it is much lower because the outside wall is so cold. As a result, the thermostat does not switch off until the wall is heated up by an excessive temperature in the room. What is happening is that the feedback system in your thermostat is getting the wrong information. The thermostat is falsely told that the temperature in the apartment is lower than it really is.

That is similar to what happens to the fetus when it is exposed to excessive maternal cortisol in the womb. When high levels of cortisol leak through the placental barrier, the fetal brain creates fewer cortisol receptors in the region of the brain that senses and regulates the amount of cortisol in the blood. After birth, everything is fine until stress strikes. Then, just like our heating system analogy, the cortisol-producing system swings into full gear, but the desensitized cortisol feedback system doesn't turn down the stress response until it reaches inappropriately high levels. In our analogy of the apartment heater, we could solve the whole problem by putting the thermostat in a better place, an inside wall for example. However, in real life the cortisol feedback system is stuck that way forever.

The result of this resetting of the stress axis is a tendency to higher cortisol levels and an overactive stress response system throughout life. It is possible that this is an adaptive response. In other words, perhaps Mother Nature allows alteration of a child's stress response system because it makes a child more likely to survive in some situations. We've seen how a fetus that is malnourished in the womb is metabolically programmed to latch onto calories and store them away as fat. This adaptive response promotes the child's survival if lack of nutrition continues to be a problem after birth. Similarly, if a child is being born into a dangerous world, perhaps a stress response that has been heightened prenatally would serve to make the person more alert, more sensitive to danger signals, and quicker to act when threatened. But when the world is not that dangerous, then a heightened stress response can be inappropriate and detrimental. The child might grow into someone who flies into a rage over getting the wrong order at a restaurant, or plays chicken on the freeway with the guy he thinks cut him off.

Even when a boosted stress response aids survival in the short term—in the face of danger—a permanently overactive stress system is clearly always harmful in the long term. Higher cortisol levels weaken many kinds of cells, and the link between stress and cardiovascular disease has been much discussed over the last three decades. But many other organ systems and cells are susceptible to injury by high cortisol levels. Cells exposed to chronic stress will be susceptible to premature aging and death. In extremely and excessively stressful situations, degenerative changes can occur in a matter of weeks. This is especially true for nerve cells. Excessive amounts of glucocorticoids in the bloodstream have been linked especially to accelerated aging in the hippocampus, the part of the brain responsible for memory formation and storage. In contrast, if chronic glucocorticoid exposure in rats is reduced, the animals live longer and age more slowly. Scientists know that humans under stress can also be troubled by memory problems.

Not only are people with an altered stress response harmed by the high cortisol levels in the blood, they are more likely to become stressed—to attain those harmful levels of cortisol—than the normal population. In other words, the awful nature of this problem is that for a given stress—the loss of a job, divorce, problems with a wayward child—people with a muted cortisol feedback system will overreact, having a greater "flight or fight" response than is appropriate or helpful in normal society.

The programming of the fetal stress response system happens in different ways throughout pregnancy, but stress probably causes most problems during the second and third trimester. The second trimester is when the brain is being shaped in earnest, and key layers of gray matter in the cortex are being established. During the third trimester the fetus is preparing in earnest for life outside the womb, getting the stress response system ready for the very stressful process of labor and birth.

THE STRESS-DEPRESSION LINK

Of course, there is another possible response to stress than "fight or flight." Someone faced with a chronic stressor could just give up. There is a great body of evidence that part of the cause of many cases of clinical depression involve the body's stress response. Animal research has shown that animals faced with a chronic, inescapable stress will develop all the external signs of depression—lethargy, loss of appetite, decreased sexual drive, and fearfulness. People who have been exposed to long-standing psychological trauma also can develop signs and symptoms of depression. It is almost as if the brain, when faced with a battle ,

If chronic stress is involved in depression, we would expect to see an altered stress response in clinically depressed individuals—which is exactly what we do see. Many studies have shown that the brain-pituitary-adrenal stress response is altered in people afflicted by this problem. Depressed individuals have a higher level of cortisol in their bloodstream throughout the day.

They also have another distinct alteration of the cortisol response system. Cortisol levels in the blood naturally fluctuate in all of us throughout the day, going through a rhymthic rise in cortisol early in the morning as you prepare to awaken and reaching a peak in the earliest hours of wakefulness. The levels of cortisol fall naturally throughout the day and reach a low point in the middle of the night, during sleep. This fluctuation of the levels of cortisol in the blood is one of our most basic biological rhythms. Variations in these biological rhythms can greatly affect one's mood. Some people do not like to be sociable in the morning. They may be the life of the party later in the day, but in the morning their internal clock makes them moody and withdrawn. In contrast, if you live with someone who is a true early bird, you are advised to make allowances for their proclivities.

In depressed people this daily ebb and flow of cortisol is blunted. The constantly high levels of cortisol and a lack of daily rhythm in terms of cortisol fluctuation seem to indicate that the stress response is always on, mirroring the kinds of problems we see in experimental animals that have been subjected to chronic stress. It is almost as if the brain, which is usually spurred into greater activity by stressors, can't maintain an appropriate response to the constant barrage of stress indicators and begins to shut down, to turn away from the world.

All of these observations indicate that people who are prone to overreact to stress, whether as a result of prenatal environment, postnatal experience, or genes, are more likely to become depressed than those who don't live with this burden. Of course, we can't confirm that this is so, or know why, until we discover the exact mechanisms of depression and understand precisely how it develops. What we can say is that genes are not the sole determinant of how we think and what we feel. Both the postnatal and prenatal environments play a major role.

One highly interesting—and important—question for the future is this: can exaggerated stress responses be passed on across multiple generations? I've talked about how other prenatally influenced conditions, like diabetes, can pass from mother to daughter, and from daughter to grandchild. It's possible that a highly stressful pregnancy for a mother may alter her daughter's stress response in such a way that the daughter's stress levels remain high throughout life—and make it much more likely that unless stress levels are reduced to acceptable levels, the daughter's own pregnancy will be stressful too. If so, the grandchild may have a lifelong tendency to overreact to stress, even if he or she is brought up in nonstressful surroundings. So there is a great benefit to understanding, and doing your best to control, stress in your life.

STRESSFUL NEWS?

I think that first we have to acknowledge that some of this new scientific information can be disturbing, and perhaps even seem unfair. Mothers have long been unfairly burdened with responsibility for every problem their children might face, and now they are being told that the burden begins during pregnancy. They might feel that they just cannot win.

Let's agree, though, that it does no good to ignore the realities of the importance of pregnancy and its potential to give the gift of good health or less-good health. We have to accept compelling new information as we come to it, and incorporate it into our lives. At one time it was news that women should abstain from drinking and smoking. Many women were angry at being told they had to break their strong addiction to cigarettes, or felt juvenilized by having to drink nonalcoholic drinks at cocktail parties or other social occasions. Now the situation is quite different. Most pregnant women accept totally that they shouldn't drink or smoke during pregnancy, and see it as a small burden to take on for the sake of a healthy baby.

Similarly, the idea that you have to avoid excessive stress while pregnant may seem at first to be another unfair burden. But in my experience, once you accept that a lower-stress pregnancy is essential for a healthy baby, you will be able to give yourself permission to make changes to lower stress in your life.

What do I mean by this? I mean that, like many women, you may find it hard to do something that benefits you and you alone. You might therefore not allow yourself to take it any easier during pregnancy. You might feel you have to keep on top of a full load of responsibilities at home and at work despite the extra physical stress that pregnancy puts on your body, and then let yourself take a break only when it is physically or psychologically impossible to go on. But once you understand that identifying and alleviating the sources of stress in your life is not a selfish act, but something you are doing for your child's health, you may be more able to give yourself permission to cut back.

The first principle is to be good to yourself and your child. Have others around you take some of the burden off your shoulders. We all tend to feel that things will fall apart without our full efforts, but we usually discover that we can cut ourselves a little slack if we plan for it and if the reason is important enough. This new perspective should spur you to take a few simple actions to better manage your stress. As one of my close friends often says, pain is usually inevitable but stress is, to some extent, optional.

CONTROLLING YOUR STRESS RESPONSE

I hope you now see how important it is to control your response to stress during pregnancy, but how do you accomplish this? Pregnancy is an unknown and uncharted sea the first time you set sail on it. Even when you have been through it once or twice before, no two pregnancies are alike. Even if you have close support from your partner and extended family, there will always be new difficulties to face. It's hard enough to keep yourself healthy and take care of normal, everyday responsibilities without some elevation in cortisol levels. If you also work outside the home, or have other children, staying calm and stress-free while pregnant might seem nearly impossible. Even thinking about the need to reduce stress can be anxiety-producing. But the fact is, you can always reduce stress. Always.

Undoubtedly you face myriad challenges every day which have the potential to add stress to your life. However, although these events are *stressors*, they do not necessarily have to lead to stress. Hard as you try, you may be unable to change all the difficulties that life and pregnancy throw at you, *but you can change your reaction to them.*

There are concrete techniques that you can use to insulate yourself and your baby from the effects of the stress response. A huge body of research and thousands of years of experience show that how we react to problems can be modified so that the problems don't bother us as much. You *can* make your life less stressful. Not stress-free—but stressed less.

At the same time that you are making your own life less stressful and lowering the amount of stress hormones that your baby is exposed to, you will be making your own life more enjoyable. Pregnancy should be a special time in your life, a time when you grow in spirit while nurturing a new life within. As with any worthwhile growth experience, there will be challenges and difficulties. But you should make the most of this special time. A suffocating feeling of stress, the feeling of being overwhelmed by life, is the surest way to snuff out the joy of this marvelous experience. Organizing your life in order to minimize stress is therefore one of the most important things you can do during your pregnancy, both for your baby and for yourself.

The first step toward lowering the amount of stress in your life is to identify the major sources of stress you experience. Is it work outside the home? Finances? Being a single mom? What part of your daily life most needs addressing? Of course, it probably will not be easy to change these things. If you work outside the home, it may not seem beneficial to your career to take a leave of absence or work part-time. But keep in mind that pregnancy doesn't last forever, while the effects of a negative womb environment can last a lifetime. If people in your life understand the real effects of stress, they will be more inclined to support the changes that you need to make together.

Remember that pregnancy is a time of great change. It represents more of a "change of life" than menopause. Your internal organs and neural circuits are being rearranged. Your internal world and your sense of self are undergoing a revolution. That internal change needs to be mirrored by external change. You need to take advantage of the revolution in your sense of self and your relationship to the world to reorganize your life. In many ways, that reorganization will happen anyway, so you're better off if you think clearly about what you truly want and need in life. And the changes you initiate should start right away, as soon as the internal changes of pregnancy begin. Don't wait until you are "showing" to rearrange your life to reduce stress.

By learning to control stress, you will also be picking up skills that you will continue to need as you raise children. The potential to feel stressed never goes away. If we are smart or lucky, however, we keep learning how to manage it. Remember that the biggest difference between children and adults is not that we know more facts—a child genius is still a child—but that we adults can learn how to manage our emotions. It's something we all should keep learning, every day.

In order to reduce the number of stressors in your life, get others to take on some of the burden that you have been carrying. Share the problems. Evaluate what is essential in your life and what is not, and discard or put off nonessential duties. Surely we know enough about the importance of pregnancy, even before the new knowledge presented here, to understand that for a few days a week your house can be less tidy than usual. You may need to ask your spouse to take over many of your normal responsibilities. After all, producing a new baby is a job you are doing for the whole family, twenty-four hours a day. The women in my family all took an afternoon siesta every day from about the halfway point of pregnancy, even though this was not easy for others in the family. I know that not everyone can manage this, but it is a goal that you can strive for.

The next step is to reduce your response to stressors. When you can't change the problem, change your reaction to it. Countless studies on stress show that the major recipe for protecting the body from stress can be reduced to a few simple elements: knowing when a threat is serious or not, taking control of the environment where one can, seeking appropriate outlets for frustration, viewing bad events as isolated incidences, and seeking social support. When stressful situations have a large psychological component, stress is very manageable.

There is a large body of research on successful stress control.

What follows here are short descriptions of a number of different techniques that have been proven to fight stress effectively. Of course, many books have been written about each of these techniques, and to get the most out of any of these methods you should consult one of these or take a class.

One of the most important commitments you can make to yourself is to set aside some time every day for your stress-relieving activity, even if you have a million things to do, and even if all of them seem so much more important than sitting quietly and thinking about your breathing. Remember that your stress relief activity takes only a few minutes out of the day, but can bring immediate benefit to both you and your baby, decreasing the stress both of you feel throughout the day. One character in the movie *Broadcast News* relieved her stress by scheduling a few minutes of a good cry every day—funny, but not a bad idea. Whatever else you have planned, schedule some time for easing up on the stress-response accelerator.

STRESS-FIGHTING TECHNIQUES

I have highlighted a few of the most popular techniques to give you a start in controlling your stress, but I encourage you to learn more and find a technique that works for you.

EXERCISE

Long ago, before the dawn of modern civilization, most of life's threats were physical. The stress early humans felt was mostly the result of the threat of physical attack by a predator or by another member of the social group. For physical threats, the "fight or flight" response is aptly named—physical threats demand a physical response. In modern times, we are fairly safe physically, and our main stressors are complex and more psychological in nature. We worry about being able to juggle bill payments successfully. We worry that we are letting down a friend, or that a friend is mysteriously turning cold toward us. These are mental threats, but they can feel just as real biochemically as direct physical threats to our bodies. Unfortunately, millions of years of biological conditioning don't fall away in just one lifetime, and our bodies still respond accordingly. Our heart rate and blood pressure still rise, we get jumpy and ready for fighting or fleeing.

Exercise can be an excellent way to resolve the physical portion of the stress response. By releasing some of that physical tension, you are also turning down the stress response in the brain, allowing you to feel less on edge psychologically. Since exercise has so many positive effects, and since the subject is fairly complex, I am devoting all of chapter 5 to its discussion and will say no more about it here.

PROGRESSIVE RELAXATION

Like exercise, progressive relaxation works on muscle tension, one of the natural components of the stress response. You might think that muscle tension is merely produced by stress, and that if you get rid of the tension you are still left with the other elements of the stress response—the troubled thoughts, the worry. The odd thing, though, is that psychological and physical reactions to stress are intertwined. By removing the muscle tension, you will actually feel better psychologically. The link between mind and body is a two-way street. The state of the brain will affect how muscles react, but the reaction of muscles will also affect the state of the brain. The most interesting illustration of this was an experiment in which researchers tried to find out if smiling could make people happier. What they found was that if the facial muscles that produce a smile are made to contract with a mild electric current, people often report feeling happier than if these muscles weren't stimulated, or if stimulation was administered to other muscles.

Progressive relaxation is a way of focusing on each of the major muscle groups to identify where tension lies, and then relaxing those muscles. It also teaches you to recognize what tension (and relaxation) feels like in each of these muscle groups so that you will be able to recognize building tension during the rest of the day. You can do progressive relaxation either lying down or in a sitting position with your head supported by a pillow. Really, the only restriction is that you should be able to relax completely without falling over.

Start by clenching your fists. Notice how this muscle tension feels. Keep your fists clenched tightly for five to fifteen seconds, or long enough to remember what this muscular tension feels like. Then relax your hands. Note the limpness of the muscle in your hands. Next, tense your arms—your forearm muscles, your biceps and triceps. Hold that position until you feel the strong tension, then relax and let the tension flow out of you. Experience how flaccid and heavy your muscles feel.

Work on your head next. Tense your forehead and squint your eyes tightly. This is the worry position. Feel the tightness around your eyes and the wrinkled brow. Hold this position, noticing exactly how this feels. Does it feel familiar? Is this the position your forehead is in much of the day? Then let forehead and eyes relax. Notice the looseness of muscles in the smooth brow, the heavy feeling around the eyes. Repeat this sequence with the muscles in the mouth and jaw, and then repeat with the neck.

Next, take in a large breath, hold it, and tense up all the muscles in the chest and shoulders. How often does this feeling occur for you? It's as though you are just about to get bad news. After learning and remembering this sensation, let your breath out slowly and allow your muscles to fall limp. Feel the heaviness in your chest. Draw in another breath and exhale slowly, letting even more muscle tension go. Slowly inhale and exhale a few more times, with each breath ratcheting out more muscle tension.

Now tighten the muscles in your abdominal wall. This won't cause labor contractions to begin, but you might activate a few more Braxton-Hicks contractions (the uterine hugs for your baby) than you usually would. Hold this, then slowly release all the tension, and let it flow out of your stomach muscles. Lastly, tense your buttocks and legs. Hold tightly. Imprint the memory of this sensation on your brain, and then let all the tension flow out of these muscles. Notice how floppy and relaxed the muscles in your legs are.

Now let this relaxation spread up through your abdomen, your chest and shoulders, your face, your arms and hands. Mentally check each area of your body and let more and more muscle tension out, even when you think that area is already relaxed. Lie or sit like this for a few minutes and let your body learn what this feels like—absorb the whole feeling.

Progressive relaxation, like most relaxation techniques, is something that you can get better at through practice. It takes about a week or two to learn to do it well, and with further practice you can accomplish a fairly deep relaxation much more quickly than when first learning it.

MEDITATION

For many Americans, meditation has the image of some sort of flaky hocus-pocus for people dabbling in Eastern religions. But anyone who has actually looked into meditation, including medical researchers at Harvard and other leading academic institutions, knows that meditation is fundamentally a nonreligious way to calm the body and mind. Meditation doesn't have anything to do with tying your body in knots, or lying on a bed of nails, or walking on coals. Meditation is simply the practice of uncritically focusing the mind on one thing at a time.

What does this do? Well, it is impossible to feel anxiety and stress when your mind is not considering the people, events, or emotions that are making you feel stressed out and tense. Most of us have all sorts of thoughts that continuously flit through our minds. Some thoughts pass in and out of our consciousness so quickly that we may not be aware of them, and yet these unacknowledged ideas or feelings can produce generalized anxiety. Many people are going about their job or housework, yet their brain is working on all sorts of other problems. With all these problems constantly circling around the mind, like airplanes stacked up over a fogged-in airport, it is impossible to feel calm and in control.

Meditation is a method for clearing out all those circling thoughts. When this happens, you also rid yourself of the anxiety that is often keeping you from solving problems. In addition, when you practice meditation regularly, you find that you have more control over what you are thinking and feeling. Furthermore, when you rid your mind of anxiety, you will be able to think more clearly about a problem and come more easily to a solution. It's a little like turning off the sound during a horror movie. You may not be thinking about the scary music, but it affects how you feel about what you are seeing. But if you turn off the sound, the movie does not seem so threatening.

Once you are able to control the kinds of thoughts you have, you will find that it is easier to identify various thoughts and associated sources of anxiety that are all jumbled together. The jumble of thoughts is often what keeps people from knowing what they want, or knowing what is the right thing to do. When you are able to separate and consider each source of anxiety separately, your choices become clearer.

There are many different kinds of meditation, though the goal of each is the same: a clear mind, a relaxed and stress-free body.

Although people get better with practice, the basic meditation method can be learned in minutes. The essential elements of meditation are

1. A relatively quiet place. It doesn't have to be perfectly quiet to meditate. In fact, with practice you could meditate right in the middle of a World Series game. But for beginners, quiet helps you focus. An ideal place might be an empty room that you have set up as a nursery. The nursery is usually the quietest room in the house, and it might help calm you to think about your baby taking a nap there in the future. There are always some sounds around, however, so don't obsess about eliminating every little noise before you get down to business.

2. A comfortable position. For beginners, sitting in a chair with hands resting lightly on your legs is fine. This is a position most of us are used to, so it is comfortable not only physically but emotionally—no need to be embarrassed about sitting in a "weird" way. Some people prefer to sit on the floor Japanese-style, kneeling and sitting back on your feet, with toes together and knees either together or apart, and hands on the

legs. The stereotypical meditation position, cross-legged with both feet resting on the tops of the thighs (the full lotus) is too difficult for beginners and is certainly not necessary for meditation. In all positions, the back should be straight to make breathing easier. This is particularly important in the last stages of pregnancy, when you will need every bit of help you can get to breathe easy.

3. An uncritical (passive) state of mind. Make sure that during the time you have set aside for meditating you have no other responsibilities. You are there simply to meditate, so whatever happens during that time is what happens. Don't worry that you are not doing things correctly.

4. A mental device for focusing the mind. This is simply a thought, word, or visual object that can occupy your attention. You might focus on a specific name for your baby, or on an image of what your baby looks like right now.

When you have found your spot and are seated comfortably, allow your body to relax. Try to attain an uncritical state of mind—you are just going to try this and see what happens. Take a few deep breaths. Think about your breathing, how the air feels going in and out, and whether you are breathing from your chest or your gut. Try both ways of breathing. There are many different mental devices you can use to focus the mind. To start with, pick one of the following:

Breath counting. After sitting, relaxing, and taking deep breaths, start counting your breaths up to three and then start at one again. If you forget where you are in the count or go beyond three, that's fine. Just start at one again. If a stray thought comes to you, just note it and let it go. You might be startled by a loud noise, or interrupted by some other sound. That's okay. Just note the disruption and return to counting.

Gazing meditation. Look at a small object or image in front of you. Allow your face to relax. Allow your eyes to explore the object, as if it has come from another world and this is the first time you have seen it.

Mantra meditation. This type of meditation also suffers needlessly from stereotype. A mantra is simply a word (real or nonsense) that is repeated over and over during meditation. The mantra can be expressed aloud or silently. Many like to chant *om* during meditation, but any word will do. You might try silently repeating *pickle* or *umbrella*. Any word (or baby name) you like will do the trick.

During any of these forms of meditation, other thoughts will intrude. A typical beginner's session using *Peter* as a mantra may go like this: Peter, Peter, Peter . . . I really need to drop off the dry cleaning . . . Peter, Peter . . . I'm going to have to work late to finish that project by the weekend . . . Peter, Peter, Peter, Peter, Peter, what am I going to make for dinner, Peter, Peter, Peter, Peter . . . why am I wasting my time when I have so many things to do? . . . Peter, Peter, Peter, Peter, Peter, Peter, Peter, Peter . . . I could really go for a glass of water right now . . . Peter, Peter, Peter, Peter . . . Peter is a really nice name . . . Peter, Peter, Peter, Peter.

Whenever a thought intrudes, note it and go back to your baby name, word, counting, or gazing. One thing you will realize is how many intrusive thoughts you regularly have rooting through your mind during the day. You might even find that one of these thoughts is something you didn't even consciously realize, something that has been bothering you that you simply pushed out of your mind. Set it aside and resolve to return to it when meditation is over.

Meditation is an excellent source of stress reduction and a good coping skill generally. I highly recommend that you try it and learn more about it.

YOGA

Somewhat related to meditation is yoga. In a broad sense, yoga is an Indian philosophical system, but in the West one particular school of yoga, called Hatha yoga, has become popular as a form of exercise and relaxation. Hatha yoga teaches certain postures (called asanas) and breathing control as a way of relaxing the mind and body. The most commonly recognized yoga position is the lotus position, a sitting position in which the legs are crossed so that each foot is on the opposite thigh. But yoga teaches many other positions, such as simply lying down. Since there are so many different schools of yoga, some of which emphasize diet and philosophy as well as relaxation, I won't recommend a particular form here. There are many good books on yoga, and classes in many communities.

THE GIFT OF LESS STRESS

These are only a few of the methods that people use to release the constricting bands of stress and tension. Although it is impossible to completely obliterate stress during pregnancy, or at any other time in your life, the important thing is to learn how to avoid and manage it. Keeping a lid on stress and putting yourself in a calmer state of mind can make your child more balanced when she grows up, more able to feel that life is manageable, more able to be happy. And as with so much else in parenting, the gift of a more relaxed state of mind will come back as a gift to yourself.

5

Exercising for Two: Reducing Stress and Augmenting Health

No pregnancy issue has been more debated, and based on less information, than the question of women's physical activity during pregnancy. Happily, there are a few researchers (some of whom are also practicing obstetricians) who in recent years have devoted themselves to finding answers on this subject. What they have found is that, contrary to the advice of grandmothers and much of the medical establishment, moderate exercise within the range you are accustomed to is extremely helpful to both mother and child. In fact, they are finding that

- A mother who is herself in good shape creates a healthier newborn with higher initial scores on tests of physical well-being.

- Exercise helps mothers cope with stress and lessens mood swings.

- Exercise helps lessen many of the bothersome physical side effects of pregnancy.

- Exercise makes labor and delivery easier and faster.

Just as by practicing good nutrition and stress relief, by staying in shape and continuing the appropriate amount of exercise during pregnancy, you can be good to yourself and your baby. Everything about exercise improves how you feel about yourself and your body during a time of extraordinary change. At the same time, exercise works with the principles of good nutrition to create a healthy environment for fetal growth.

And by exercise we are not just talking about a few knee bends. Research is showing that pregnant women can tolerate a good, solid workout at 50 percent of their prepregnancy capacity. You may have to change the kinds of exercise you do or adjust the way you work out to accommodate your changing body, but you are encouraged to continue appropriate levels of exercise while expecting.

EXERCISE DURING PREGNANCY, HISTORICALLY

Why should the idea that exercise is okay, even good, be such amazing news to us? Why haven't we accepted these ideas all along? After all, pregnancy is a normal part of life. Pregnant animals continue to hunt, forage, make nests, defend their territory, and do all the normal labors of life during the gestational period. Historically, the human animal was required to continue the hard physical labors of life while pregnant. Why would anything as normal and necessary to life as exercise be threatening to healthy pregnancies and healthy babies?

The answer lies in more recent history. During the Victorian period, society felt so affronted by anything hinting of sexuality that people were encouraged to refer to table "legs" as "limbs." Pregnancy was an all-too-visible reminder that a woman had performed the most intimate act of all, and so women in the "respectable classes" were often taken out of the public eye and put in "seclusion" as soon as they were visibly pregnant.

A positive development during the same period was an increased concern with, and understanding of, disease and hygiene. Along with a general interest in public health, there was a new interest in healthy babies and in keeping mothers healthy during the prenatal period. Unfortunately, although this interest was generally good for babies, it also led people to become concerned, in misguided ways, about women's health. During the nineteenth and much of the twentieth century, vigorous exercise was viewed as one potential risk. Doctors thought exercise was connected to low birth weight and miscarriage.

Indeed, the French obstetrician-researcher Dr. Emile Papiernik traces the concern about exertion during pregnancy to studies done in 1895 by Adolph Pinard on girls in the laundries of Paris. He found that the laundresses who continued in their heavy, physical work throughout pregnancy were more likely to deliver prematurely and their babies were more likely to be less healthy. But new studies have shown that, for whatever reasons, there is a huge difference between extreme amounts of repetitive labor (often very heavy labor such as lifting heavy wet sheets) performed as part of a job, and exertion done as part of a controlled exercise program. Exertion undertaken as part of an exercise program, without the psychological stress associated with tedious and unrelenting work, does not cause an increase in prematurity.

Doctors were also influenced by some tests done on nonpregnant animals and men showing that exercise produced a sharp rise in internal temperature and a sharp decrease in blood flow to internal organs. Since the researchers knew well that high temperatures and a shortage of maternal blood to the placenta could be harmful to the developing fetus, they surmised that exercise during pregnancy was probably harmful.

Another link in the chain that kept pregnant women from exercise was found in the 1960s, when scientists uncovered the connection between psychological stress in the workplace and difficult pregnancies, and assumed that the physical stress of exercise must be similarly harmful. Then there was the belief that some miscarriages must be caused by exercise. It seemed intuitive that an embryo could "shake lose" from the uterine wall if jostled, or that muscular contraction in the abdomen might push the fetus out. The fear of starting off labor prematurely probably originates from the incorrect view (held by many) that the voluntary muscles of the abdomen are the force that causes delivery during labor. When a miscarriage occurred, it was natural to look back at what happened before the miscarriage and decide that some easily identifiable event, such as a bout of unaccustomed exercise, had caused the unwanted early delivery. So if a woman was doing anything out of the ordinary, like exercising, or even doing too much of an ordinary activity like climbing stairs, then that exercise or activity must be the cause. We now know that perhaps 30 percent of pregnancies end in miscarriage, most often for reasons that we can only guess at, so a small proportion of women who exercise are bound to have miscarriages that are totally unconnected to their physical activity. To date there is no association of miscarriage and excerise that has been shown clearly to everyone's satisfaction.

With all these ideas and conjectures floating around, the medical establishment took a very conservative position on the subject of maternal exercise and recommended against it. Over time, the suppositions about exercise being bad for pregnancy became part of the body of accepted medical fact. There are many examples in medicine where myth all too easily becomes fact. Repetition often confers a degree of authenticity on unverified notions, especially when the ideas are repeated often enough. But where was the clinical data that proved these points? The answer is that there was none.

This was the state of affairs all the way through the mid-twentieth century, until both men and women launched a health and fitness revolution in the 1970s. Women jogged, Jazzercized, swam laps, bicycled—and were delighted when they discovered that it felt good. Women felt empowered by their bodies, and when they became pregnant many wanted to go against the ideology of their doctors and their mothers and keep exercising. Many of them just felt so much better when they did. So women started asking their doctors if it was okay to keep working out while expecting.

One of the leading obstetricians who listened and thought hard about this issue was James F. Clapp III, who practices and does research at MetroHealth Hospital in Cleveland, Ohio. In 1976, a patient questioned Dr. Clapp about whether it was safe to keep jogging while pregnant. Not knowing the answer, Clapp went to the medical literature to seek one. He was unable to find any firm information in the scientific journals on which to base his answer to his patient because no one had ever researched the topic. Rather than let the matter rest there, Clapp decided to find some real data on the question of exercise during pregnancy. What I am going to tell you in this chapter draws on work from many researchers, but I would like to acknowledge a large debt to the research and writings of Dr. Clapp, including his wonderful book, *Exercising through Your Pregnancy*. Dr. Clapp was the first to really dig into this issue, and in maintaining such dogged pursuit of these questions over

Do ADD Kids have elevated cortisol?
Were their mothers stressed during pg?

the years, he has been a veritable truffle hound for truth about exercise and pregnancy.

FANTASTIC FINDINGS

In the late 1970s, Clapp obtained funding from the National Institutes of Health that enabled him to start testing the effect of strenuous exercise on pregnant sheep. As is often the case, he needed to start with studies in pregnant animals to understand the biology better before he started his research with pregnant women. He put pregnant ewes on a treadmill while he measured their vital signs—heart rate, blood pressure, etc.—as well as those of the fetal lamb they carried. Clapp tells an amusing story about the first time he did this experiment: he was amazed to find that as he increased the treadmill speed the ewe didn't seem to increase her respiration or heart rate at all. Even as he raised the treadmill speed to a dead run he could still look over and see the sheep happily looking over the wooden enclosure that housed the treadmill and chomping on grain. At that point he went over to see what was happening, and found that the ewe had braced her feet on the stationary metal strips that ran along both sides of the treadmill. After Clapp fixed the treadmill so the sheep couldn't cheat, he slowly pushed the ewes to their maximum trotting speed and searched for signs of trouble in the fetus. Despite the fact that the internal body temperature of the ewe increased and blood flow to the placenta decreased, Clapp found no signs of fetal distress. The ewes were exercised regularly and yet had normal pregnancies and gave birth to normal lambs. In later years, other researchers undertook similar experiments and confirmed these findings. Since Clapp's initial experiments were quick, he wasn't able to look closely for differences in how the lambs fared as they grew over the long term, but he found these research results enormously reassuring and for a time turned his attention to other research projects.

But sheep are not humans, and Clapp realized that to find out if people were similarly unaffected he would have to follow up these reassuring experimental studies with investigations in normal pregnant women. In the mid-1980s he began to feel it was important to settle the issue. Several clinical studies had shown that a large percentage of women who exercised regularly planned to continue exercising after becoming pregnant. And yet the American College of Obstetricians and Gynecologists, the guiding body of the profession, still recommended against exercise during pregnancy, advice that was echoed by most doctors. The result was that women were not telling their doctors that they were exercising. One study at Clapp's hospital revealed that none of the ninety-six pregnant women in the study who exercised three times a week or more ever mentioned that fact to their doctors. And the doctors didn't ask. However, if exercise in expectant women was harmful, Clapp wanted to find out.

THE BENEFITS OF EXERCISE

For both men and nonpregnant women, the benefits of exercise are well established. Exercise has been shown to increase blood supply, lower blood pressure, enlarge the heart's chambers, and increase maximum cardiac output. Regular exercise also fosters the growth of new blood vessels and increases the cell's metabolic efficiency, not least by improving delivery of oxygen to cells and increasing the number of energy-producing mitochondria in each cell. In addition, regular exercise improves the body's ability to lose heat through the skin, via both increased capacity for sweating and more rapid and extensive dilation of the blood vessels near the body's surface.

Does this sound familiar? It should—these changes are very similar to the natural changes that take place in a woman's body when she gets pregnant. When pregnancy occurs, there is a very quick increase in blood volume, a hot sensation as blood vessels near the skin flush, an increased tendency to sweat, and an increase in oxygen delivery to tissues. It seems as if exercise and pregnancy put the body through similar changes, and that exercise should work with and help the changes that pregnancy naturally fosters. That is what Dr. Clapp thought, and in large part that is what he discovered. But that is not the whole story.

What he found on further research is that pregnancy not only changes the state of a woman's body at rest, but that pregnancy radically alters how her body reacts to exertion. In the first group of women that Dr. Clapp tested, he found that women's heart rates during exercise went sky-high, far higher than they had ever gone before during the same exercise routines. Instead of the usual 150 beats per minute that they might experience during aerobic exercise, they could have heart rates of 180 beats per minute. The women themselves had noticed this and were concerned about it, asking if the increase was normal, and safe. Clapp also found that internal temperature is lower in pregnant women both at rest and during exercise. Finally Clapp discovered another amazing change: after a period of exercise, blood sugar falls— exactly the opposite of what occurs when nonpregnant women and men work out.

In short, pregnancy alters a woman's reaction to physical

exertion in a way that protects her fetus. More sweating helps a pregnant woman to lose more heat and results in a lower core temperature. Higher heart rates mean a greater flow of blood through the body to carry heat to her body surface, as well as a richer supply of oxygen to body tissues. These adaptations are probably the reason that babies born to women who do exercise while pregnant are normal, and why the lambs in Dr. Clapp's first experiments showed no ill effects from their mothers' treadmill work.

If you are exercising regularly and vigorously before pregnancy begins, you should already have increased your blood volume and developed a stronger diaphragm—the muscle that draws air in and out of your lungs. This gives you a greater aerobic capacity. Your body will already be well along the path that all women's bodies take when pregnancy begins. The right amount of exercise during pregnancy will continue this preparation because the pregnancy and the exercise are working together. In fact, not only does exercise help prepare the body for pregnancy, but pregnancy can help prepare the body for exertion. Many professional athletes have noted that they have greatly improved performance in the year after having a baby. In these women, the pregnancy acted as a good sustained training, increasing their cardiac output and aerobic efficiency. The increase in blood volume they experience as a natural part of pregnancy also gives them the same benefits conferred by blood doping—the illegal (in international competition) practice of withdrawing and banking your own blood and then putting the blood back in the body a short time before a race in order to increase your body's oxygen-carrying capacity.

BENEFITS FOR BABY

This increase in maternal blood volume and blood vessel formation directly benefits the fetus because it fosters the growth of the
placenta and allows it to work more efficiently. In addition,
maternal exercise appears to increase the baby's own blood supply
and may foster the development of new fetal tissue.

Perhaps more importantly, when a mother exercises, she
alters the environment in the womb and provides a small, controlled physical challenge to her baby. Biological systems need
periodic physiological challenges to be able to respond rapidly
and efficiently when larger stresses come along. We exercise so
that when we have to climb a long flight of stairs or run for a bus
in cold weather we don't get winded. When a mother exercises,
she is not only changing her own heart rate, blood sugar level, and
blood oxygenation, she is also changing those values for the fetus.
These fluctuations seem to make the baby able to better handle
the natural stresses of birth and of adjusting to a new food supply
as a newborn. In addition, exercise produces myriad sensory stim-
ulants for the fetus, who feels increased motion, vibration, and
new sounds. There has been a fair amount of research demon-
strating that stimulation in the womb affects the development of
the fetal brain and body. We know, for example, that the periodic
squeezes the fetus gets from Braxton-Hicks contractions alter the
sleeping patterns of the baby in the womb.

Are any of these effects measurable? Dr. Clapp and others
have measured the physical and psychological well-being of
newborns whose mothers exercised during pregnancy, and been
pleasantly surprised. Babies of mothers who exercise seem to
handle the stress of birth better than babies of mothers who
don't. The babies whose mothers exercised have fewer inci-

dences of meconium in the amniotic fluid, which is a sure sign of fetal stress. Some doctors were concerned that the motion of exercising might lead to more incidents of the umbilical cord wrapping around the neck during birth, but such incidents were actually a little lower. Levels of a key blood-producing hormone in the baby's blood were low when compared with babies of mothers who did not exercise. This may seem counterintuitive, but the levels of the hormone only rise in babies who are short of oxygen and trying to compensate by producing more blood. So paradoxically a low level of this hormone—called erythropoietin—indicates that conditions in the womb were fine and the baby did not need to increase the production of oxygen-carrying red blood cells. Most interesting was a study showing that doctors, nurses, or midwives became concerned about the baby's condition during labor about half as often when the mother had been exercising as they did when the mother had not, even when the nurses and midwives didn't know whether or not the mother had been exercising.

After birth, the babies of women who exercised were slightly leaner than those in the nonexercising control group. As we have seen in chapter 1, birth weight is extremely important as a signal of how a baby has fared during development. Dr. Barker, the original proponent of the Barker hypothesis described in chapter 1, has shown that when a baby is lighter than it should have been according to its genetic potential, there may be health troubles in store in later life. Dr. Clapp developed this idea further by suggesting that babies born of women who exercise and eat complex carbohydrates during pregnancy are a more "natural" weight, and that modern Western babies are a little heavier than desirable. Nevertheless, the difference is slight. Suffice it to say that the babies of mothers who exercise show no problems in staying warm, and have normal blood sugar levels after birth.

Dr. Clapp has found interesting psychological effects in the babies of mothers who exercise. Initially it was the exercising mothers themselves who told him that the babies seemed very easy to care for, and that they slept through the night fairly soon after they were born. Using the assessments developed by Dr. T. Berry Brazelton, Clapp found that the mother's perceptions seemed true. Clapp discovered that babies of mothers who exercise quiet themselves more easily, with less consolation from others, than babies of the nonexercising control group. The more relaxed and contented behavior of the babies of mothers who exercise may be the result of a toned-down stress response system in these babies. The behavioral tests also showed that babies whose mothers exercised during pregnancy were more aware of their environment and more responsive to it.

BENEFITS FOR MOM

I've talked a little about the general benefits of exercise, and the effects of exercise on the fetus, but there are also specific, measurable benefits for pregnant women, in addition to the few I've outlined above.

One of the immediate benefits has to do with weight gain. It is a matter of great importance that women should not try to either lose weight or retain their prepregnancy weight during pregnancy. At the same time it is not a good idea to gain too much. Scientists have found that women who exercise gain less weight during pregnancy, while still staying within the limits for a healthy weight gain.

Scientists have also found that women who exercise during pregnancy report fewer instances of discomfort due to pregnancy and during labor. In fact, labor is exactly the kind of increased stress that exercise prepares you for—through regular exercise, your body has improved its ability to supply muscles with the oxygen and glucose they need to make sustained physical effort. Labor in women who exercise is generally shorter too—a very important plus for you and your baby. Before researchers knew anything about the effect of exercise on labor, there was the fear, which became accepted fact, that strong pelvic muscles would slow down labor. A few studies have shown that on average, women who exercised for twenty minutes three times a week spent less than half the time in the "pushing" stage of labor than women who didn't exercise.

Some of the most difficult benefits to pin down are the psychological advantages that come from a regular exercise program. Women who exercise regularly during pregnancy report that they feel better, have more energy, and are more optimistic about the course of pregnancy. Is this caused by the exercise, or are women who have more energy and feel optimistic more likely to get out and exercise? There have only been a few studies in which women were randomly asked how much exercise they do (none, some, or moderate), and these studies suggest that the exercise itself does have some positive psychological benefits, regardless of a woman's initial state of mind.

THE PREGNANCY EXERCISE PROGRAM

I hope I have convinced you of the importance of exercise as part of pregnancy fitness. Now the question is, what do you do with that information? How do you put it into effect in your life to benefit your unborn baby?

What I would like to do is offer some general guidelines, some common exercises, and a few cautions about exercising while pregnant. Having said that, I may also need to point out that different people like very different forms of exercise. One person can find weight lifting enormously rewarding, while another might find it merely heavy lifting. One person may find nirvana in swimming one hundred laps, while another might find swimming laps incomparably dull. Absorb my guidelines and advice and then mold them to your own likes and dislikes.

Many of the questions that women have about exercising before, during, and after pregnancy can be boiled down to two: "How much exercise is enough?" and "How much is too much?" How much exercise do you need to do to get the physical and psychological benefits of exercise that I have described above, and how far can you go without harming your baby or endangering yourself in some way? One of the most important aspects of exercise during pregnancy is how much you are used to doing and how regularly you exercised before you became pregnant.

The answer to the first question is that there is not one hard-and-fast formula. It partly depends on what your goals are for exercising and what point of the pregnancy you are in. Exercise has different effects during different stages of pregnancy. During the first trimester and into the second, exercise can increase the growth of the placenta and the growth of the baby while decreasing the bothersome symptoms of pregnancy. During the third trimester, exercise enhances or maintains fitness, reduces weight gain, and makes labor easier. Even a little exercise early in pregnancy can enhance the natural feeling of well-being throughout pregnancy. However, it takes quite a bit more exercise later in the pregnancy to reduce any excessive weight gain and shorten labor, according to Clapp.

The main message here should be to start with what feels good. If you are already exercising, you probably know how much exercise you need to feel good. If you are just starting, you need to be aware of your own body's reactions. Earning the physical benefits of exercise in the later stage of pregnancy will take more work, although you should still do only what feels good.

What kind of exercise should you be doing? Different forms of exercise target different body parts and organ systems. Ten minutes of walking is not the same as ten minutes of bicycling. For various reasons, researchers recommend exercises in which you have to support your own weight: walking, running, dance (like Jazzercise), skating, etc. These forms of exercise confer the greatest overall benefits (walking, by the way, is generally a very underappreciated form of exercise that is easier on the bones and has great benefits). Having said that there is no hard-and-fast formula for proper exercise, I will give a general guideline that has been shown to confer benefits in research studies: twenty minutes of moderate, weight-bearing exercise, three or four times a week, seems to be a sensible target to shoot for. This is a target to work up to if you are not already doing this sort of exercise, and ideally you will start working toward this even before pregnancy begins. In addition, you should rest for at least twenty minutes after every twenty-minute workout.

The more troubling issue is how much is too much? How do you evaluate what moderate exercise is? How do you make sure the baby is not being harmed?

This can seem especially difficult because most people measure the intensity of their exercise by monitoring their heart rate, but there is some controversy about whether this is a good way for pregnant women to gauge their physical output. Remember, the whole cardiovascular system undergoes changes in adapting to pregnancy. Blood supply increases, and heart rate changes. Early in pregnancy, resting heart rate goes up as much as twenty to thirty beats per minute. Some women observe this increased heart rate in themselves and are unnecessarily frightened by it. During physical exertion, the heart rate also goes much higher than it would when a woman is not pregnant. So physical exertion that might produce a heart rate of 150 beats per minute in a nonpregnant woman may produce a heart rate of 180 beats per minute in a pregnant woman. This is entirely normal, part of the adaptations that a woman's body makes to protect and nurture her baby. Also, individual genetic differences can lead to a fifteen-to-thirty-beat-per-minute difference in people. So a single chart that claims to cover everyone may mislead people about how much they are really exerting themselves. This is yet another reason that your individual caregiver, who has the responsibility, with you, for your pregnancy, needs to be consulted about your own particular needs.

Dr. Clapp recommends that instead of using heart rate to monitor workout, you use the Borg rating of perceived exertion (RPE) scale. Because you always have some energy expenditure, even when you are totally at rest, the scale starts with a score of 6, which is what you would rate if lying in bed. A score of 11 is how most people feel while ambling along, and a score of 13 is for exertion while walking quickly. A score of 20 is maximal exertion—you have to stop or pass out.

Just as important as your own physical response is the baby's. When a pregnant woman exercises, the fetus's heart rate also increases. This is a sign of physical stress, but as we have seen, a little physical stress like exercise can be good for your baby. The point is to make sure that the fetus's heart is not beating either too fast or too slowly. The fetal heartbeat is not something you can easily check yourself, but if you are really pushing yourself hard during exercise or are trying to stay competitive as an athlete during pregnancy, it would probably be a good idea to have a health care provider monitor the baby's heartbeat after a typically strong session of exercise. James Clapp recommends an upper limit of 180 beats per minute for the fetal heart rate.

I'm sure you look forward to that first moment when you feel the baby move, and like most women you will probably find the fetus's continued movement a comforting sign (even when it causes discomfort) through the second half of pregnancy. But being aware of how the baby is moving and how those movements are changing is very important during pregnancy, and especially important during exercise. When under severe stress, the fetus's heart rate falls, and the baby will slow down or even stop moving for a time to conserve oxygen. The movements your baby makes are an important indicator of how he is feeling. You should keep in tune with your baby's activity patterns during and after exercise. Marked changes in the baby's heart rate or movements may indicate that the exercise you are doing is too extreme, or they may indicate that there is some other reason for the fetal stress and exercise merely pushes the fetus over the edge into distress. As you take your rest period after exercise, you should find that the baby moves at least every five to ten minutes (since it is easiest to feel the baby move in the last trimester, I have more details about fetal movement after exercise in that section of this chapter). Generally, you should discuss with your physician your baby's pattern of movements at different times of the day and after exercise.

THE IMPORTANCE OF FEELING GOOD

You have probably noticed that a lot of my advice about exercise centers on how you feel. That means being aware of how your body is reacting to the exercise, not only in heart rate and perceived exertion, but in other important ways:

WATCH FOR OVERHEATING

I mentioned that pregnancy changes the body's blood distribution and metabolic basis to make overheating less likely, but there is still a danger of overheating when exercise is too rigorous. The only way your baby has of losing heat is through you. The fetus in the womb cannot lose heat by the usual sweating mechanisms. So when you heat up, your baby does too. Too high an internal temperature is definitely harmful to your baby, so don't exercise to the point that you feel uncomfortably hot. If you are working up a good sweat but feel fine, you are probably okay. Increased sweating by itself does not mean that you are too hot, since pregnancy leads to increased sweating precisely to cool the body through evaporative heat loss. Still, work out in a time and place that allows the sweat to evaporate easily. The obvious example of a time to avoid would be a summer day with high humidity, when sweat doesn't evaporate well, but you should also avoid warm, enclosed gyms with little air circulation. If you have any doubts, James Clapp recommends checking your vaginal temperature right after working out, and if the temperature is below 100.4 degrees Fahrenheit at least you are not overheating.

WATCH YOUR FLUID AND SALT INTAKE

Give your body something to sweat. Make sure that you stay well hydrated throughout pregnancy but especially while working out. Sip some water occasionally while exercising rather than trying to drink a lot before or after. If you are doing a lot of sweating, you are losing salts as well as water, so drink some of any of the sports drinks that contain essential salts.

IF YOU FEEL PAIN, STOP

Pain during exercise is never a good sign. This is particularly true if you have pelvic pain during pregnancy. Back pain may be common in later pregnancy and may require a change in the way you work out. Pain doesn't necessarily mean you have to stop exercising, but it is the way your body warns you that something is not right. As with anything unusual at any time during your pregnancy, you need to consult your obstetrician.

MAKE THE EXERCISE A POSITIVE EXPERIENCE

Sticking with any exercise program takes a lot of motivation, even if you are not pregnant, and it is difficult to retain any desire to exercise if working out is a burden with few rewards except a log of tortured miles run. When nausea and fatigue beset you during early pregnancy, it is doubly important that the exercise be intrinsically rewarding. Otherwise you are likely to stop doing it, and research shows that all the benefits of exercise go away when exercise is stopped in the middle of pregnancy. A good exercise program is holistic: it works well with the rest of your life. The exercise is enjoyable, and you fit it into your schedule so that you aren't always scrambling to find time to do it. A good, holistic exercise program makes you feel better emotionally as well as physically.

Any exercise, whether for recreation or for training for serious competition, can be taken too far. Any athlete can fall prey to an overtraining syndrome. This syndrome includes fatigue, pain, susceptibility to injury, loss of motivation, and poor performance in the exercise regimen or competition. If you find yourself falling victim to any combination of these symptoms, you should cut back on exercise for a while. If the thought of cutting back sounds good, that is another indication that you may be overtraining.

DO IT RIGHT

Exercise doesn't happen automatically. It's something we need to learn how to do correctly. That learning can take place through books, via an instructor, or through experience and self-teaching. But part of making exercise feel right is doing it correctly. If you already have an exercise regimen, you will know a lot about how your body reacts to exercise, but pregnancy will change those reactions and require some new learning. If you are just starting a program or starting a new exercise, it is doubly important that you learn about what you are doing. Health club trainers are some of the best people to show you how to do each exercise properly. You might also want to consider taking an exercise class specifically geared to pregnant women.

WHEN NOT TO EXERCISE

Exercise during pregnancy is not for everyone. It is true that older guidelines have been liberalized based on research showing that for a normal pregnancy the risks are not nearly as high as the medical establishment thought. However, there are some times when exercise is still thought or known to be risky.

A series of guidellines are available from sources such as the American College of Obstetricians. In summary, you should absolutely not exercise if you have any of the following conditions:

History of miscarriages

Incompetent cervix (the cervix dilates far ahead of schedule)

Persistent bleeding after twelve weeks of pregnancy

Placental disease

Poor fetal growth

Premature labor or a history of premature labor

Pregnancy-induced hypertension

Ruptured membrane ("water" has broken)

Twins or other multiple pregnancy

You probably should not be exercising if you have any of the following, although in some cases you and your physician might be able to come up with a special exercise program.

Anemia

Breech presentation after twenty-eight weeks

Early pregnancy bleeding (more than "spotting")

Extremely overweight or underweight for your height

History of poor fetal growth

History of rapid labor

Palpitations or arrhythmia of the heart

Sedentary lifestyle prior to pregnancy

The golden rule applies—if you have any doubts, consult your physician.

Now that you understand some basics about pregnancy fitness, here are some issues for you to consider at each stage of your pregnancy.

EXERCISE IN THE FIRST TRIMESTER

The first trimester is the time when there are the fewest visible exterior changes, but when internal changes, for both the mother and the baby, are the greatest. This is the time when the placenta and embryo are growing and changing very rapidly. The first weeks of the first trimester, before many women even realize they are pregnant, is when the embryo is establishing all the organ systems that he will use for the rest of his life. During the first weeks, the placenta is also attaching itself to the wall of the womb, anchoring the baby and connecting yours and your baby's blood supply lines for taking in nutrition and getting rid of waste. For the mother, this is also the time when the vascular system is trying to rapidly increase the total blood supply, and mothers feel the classic symptoms of system "underfill"—the apparent shortage of blood leads to nausea, dizziness, racing pulse, sweating, pallid skin, and feeling too cold or too hot. All these feelings make it harder to exercise, but this is also the time when exercise can accelerate the changes that pregnancy is already inducing. Exercise also sends signals to increase the blood supply and help tissue growth, so the placenta will grow more quickly and the blood supply will increase more rapidly. If you want to get the benefits of exercise during pregnancy, now is the time to work out a good regime with your physician and your trainer, if you have one. In any event, it is important to keep up the exercise you started before pregnancy.

Of course it will be doubly hard if you are just beginning an exercise program, but start slowly and keep at it. Whether beginning or continuing exercise, you should build goals and rewards into your program. Give yourself a special snack or a present when you reach goals. Share your aims and progress with your family and friends. That helps you keep at it too. Keep a chart of your progress. This gives you an idea of how close you are getting to the goals you have set and also lets you better monitor how you are feeling while working out.

It's good to keep motivating yourself, but of course sometimes it may become too difficult to keep going. The point is *not* to stick to a program come hell or high water, but to stick to a program that feels like it is working. It's normal to feel not so great in early pregnancy, so don't let a little bit of discomfort stop you. But sometimes you may just be trying to do too much physically, and you shouldn't feel guilty about scaling back your program for a time. Try not to stop completely, but slow the program down and then try to get back to a normal schedule as soon as you feel able.

Keep in mind that the previous prohibitions to exercise apply. Stop or don't start exercising if you have:

- Vaginal bleeding and/or cramping

- Pelvic pain

- Unabating nausea and vomiting

- An acute sickness like the flu

A good aim is to try to exercise in twenty-to-thirty-minute periods, about three times a week. An upper limit on the Borg scale of exertion might be a 16 out of 20 (see the scale described above), but remember that your main limits should be self-set for you yourself, based on how you feel. Remember to take as much time to relax after exercise as you did for the exercise itself. You can do reading or other mental work during this time, just not anything very physical. It would be nice if this relaxation time could be spent doing something enjoyable like reading a good book, but it is understandable if you feel that you can't carve that kind of time out of your schedule.

Just to remind you:

- Watch your temperature and the heat and humidity of the environment you are exercising in.

- Stay hydrated. Fluid balance is important throughout pregnancy.

- Keep your blood sugar up. It's especially important for early development that blood sugar not fall too low. That means healthy snacks every three hours or so (which most women are already doing at this point to keep nausea down). You shouldn't exercise within two hours after snacking, and you should have a small snack right after exercise. As I mentioned in chapter 3, the snacks should be complex carbohydrates or fruit, since refined carbohydrates such as pasta or white bread cause the blood sugar to rise at first and then fall rapidly thirty minutes to an hour later.

Although most forms of moderate exercise serve you well early in pregnancy, remember that the experts consider weight-bearing exercise the most beneficial. Also, everyone should be doing stretching to conserve flexibility of the tendons and ligaments and to help avoid muscle injury during exercise.

Now is also a good time to start preparing for later pregnancy by doing a little of one of the exercises that are better suited to the third trimester. Exercises like the StairMaster, low-impact or step aerobics, water jogging, or swimming take some time to learn and feel comfortable with. When you are twenty-five pounds heavier you may need to change the type of exercise you do, and the transition will be easier if you have already become familiar with one of these activities.

THE SECOND AND THIRD TRIMESTERS

During the second trimester, all the fetus's organs are in place— now is the time for rapid growth. If you are worried that you are getting too much exercise during this phase of pregnancy, the best general guide is your obstetrician's assessment of fetal growth. If your regular checkups show that the fetus is growing normally, then you are doing fine on exercise. Keep a check on your baby's movements, their frequency and intensity. Every baby has his own pattern and amount of movement. If the baby is doing well, and both you and he are benefiting from your exercise, the baby's movement patterns—in both their extent and their pattern throughout the day—will be consistent.

During the third trimester, the growth of the baby naturally slows down ("Thank goodness!" is how most women respond) because there is no more space to grow. A lot of development in the third trimester focuses on getting the baby ready for life outside the womb—maturing organ systems, putting on fat as an energy bank account for the future, preparing the lungs to breathe.

The things you have to be careful of during mid- and late pregnancy are much the same as during early pregnancy, except that now your abdomen is making its presence felt. And you have to take that into account. One thing to think about is the change in your center of gravity and the possibility that you may be knocked off balance more easily while working out. What researchers find among women who exercise is actually that exercising women have a better sense of balance while pregnant because they are in good physical tune with their bodies all the time.

Because of your greatly enlarged midsection, the American College of Obstetricians and Gynecologists recommends that you don't do any exercises while lying down. James Clapp is now conducting research on whether this really is a problem, and so far he has found that it does not seem to be, except perhaps for very late in pregnancy. Late in pregnancy it actually becomes pretty much impossible to lie on your stomach, and lying on your back is not a good idea because of the possibility that the weight of the baby will compress the major vein that lies just on the back wall of your abdomen and runs along the right side of the spine, taking blood from the uterus, recycling it back to the heart and from there on to the lungs to get more oxygen. The safest way to lie is on your left side, since that takes any pressure off the vein.

Another problem late in pregnancy is that the big abdomen pushes up on the lungs and makes it harder to breathe. Exercise can help a little in this respect too, since it will make you less likely to become out of breath during regular exertions like climbing stairs. Actually, if you could see an MRI of your insides late in pregnancy, you would be shocked. Not only are the lungs left little room to expand, the heart, liver, and other internal organs are pushed out of their usual positions. One of the organs squeezed the most is the bladder, and during mid- to late pregnancy it is natural for women to urinate even more frequently than they did during the first trimester. However, don't try to drink less frequently so that you have to urinate less often. Your baby needs to fill up her body fluid spaces and develop her blood system. You must stay well hydrated, the water bottle always at hand, especially while you exercise.

In addition to staying hydrated, you should continue to:

- Keep your blood sugar up by not eating for a couple hours before exercise and eating whole grain foods or fruit right after exercise.

- Make sure you are gaining the weight you should be.

- Watch your own reaction to exercise and avoid "overtraining syndrome."

- Stay aware of how your baby is reacting to exercise. Once you are able to feel the baby kick, you should be able to feel the baby moving several times within the first twenty to thirty minutes after exercise. Remember, when a fetus is in distress, he will slow down his movements to save energy; a lack of movement may indicate that he is not happy about your working out.

As the pregnancy progresses, you should also spend more and more time resting. Rest is especially important because you will find that you expend more effort doing the things you normally do. Also, between frequent middle-of-the-night urination and the bowling ball inside your middle, you will get little undisturbed sleep. Part of the solution may be to spend more time in bed, so that ten hours of fractured sleep may give you the same rest as eight hours of good sleep. There are also various special pillows that you can wedge under your body while you sleep to give you more support, but they often fall out when you roll over. One idea that I have seen work is to make a sort of nest for yourself with a child-size beanbag chair in bed. Beanbag chairs nicely mold to your body and will remold when you move.

Staying comfortable while exercising also becomes a growing challenge as the fetus grows. Two sports bras (one on top of the other) may now be necessary instead of one. Many women find that a belt to provide abdominal support is a good idea. These belts fit around the waist, over the hips, and under the abdomen and are commercially available.

WHEN TO STOP

Many women who exercise before pregnancy and keep exercising through most of their pregnancy are convinced of the positive effects that exercise has on their body and state of mind, and for these women the important thing is holding themselves back if there are signs the exercise should stop. Among the signs that should lead to stopping exercise are:

- Pregnancy-induced hypertension.

- Rupture of the membrane ("bag of waters") surrounding the baby.

- Any recurrent bleeding from inside the womb. If this occurs, you should definitely get your physician to look into it.

- Signs that labor may start earlier than normal. Of course, this is a judgment call in many cases. If you have a history of premature labors, you will need to consult carefully with your caregiver on all your activities, nutrition, sex, stress, rest, and exercise. There are several different known causes of prematurity and many we currently know little about. Each pregnancy is unique. Some cramping is normal in late pregnancy, and exercise can make Braxton-Hicks contractions feel stronger than they might otherwise. If you feel any strong contractions, it's best to let your doctor know what's going on and let him or her advise you about whether to cut back or stop at this point.

FIT BODY, HEALTHY BABY

We've come a long way from what we knew only a generation ago about the effects of maternal exercise on the course and outcome of pregnancy. Most women are glad to find out that they can still exercise, and that the right amount of exercise is in fact good for them. As long as you monitor your progress and observe the few conditions under which exercise is still not a good idea, then you can stay in shape all the way through pregnancy, creating the finest environmental support system possible for the child inside you.

6

Drinks, Pills, and Toxic Spills

In some ways, the first months of life are the safest anyone ever experiences. Your growing child lies nested in the womb, almost weightless in the warm bath of amniotic fluid, held and protected by your body, hugged and caressed by the contractures of your womb, every need provided for. But in other ways, this is the time when the fetus is most vulnerable. Potentially toxic or challenging chemicals around us in everyday life can make their way into the womb.

Scientists are finding that:

- A fetus is especially vulnerable. Substances can be toxic or challenging to the fetus in ways that these substances are not for adults.

- Smoking is bad for the fetus in different ways during different times in the pregnancy, so quitting smoking is beneficial at any time.

- Alcohol, which is not essential for any aspect of pregnancy health, has clearly defined adverse effects on the embryo and fetus.

- Coffee causes problems in high doses during pregnancy, but seems to be fairly harmless in small amounts.

- Even some over-the-counter medications and "natural" remedies can be very toxic.

- Health foods, nutritional supplements, and herbal teas are not necessarily innocuous.

- Despite toxic threats all around us, your body has a few tricks to keep you and your baby out of harm's way.

- A preconception visit is the perfect time to explore all the items whisch can be attended to and modified to improve pregnancy outcome. These include: family history of hereditary diseases, exposures in the home, on the job or in the environment, maternal or paternal diseases or medications which may influence fertility or pregnancy health.

CLEANSE THE WOMB BEFORE PREGNANCY

Like eating right and exercising, getting toxins out of your body should start before conception. The reason for this is that some toxins can last a long time in the body. For instance, once you ingest lead or other heavy metals, they can stay in your body for months, finding refuge in fat cells or bone tissue. If your house or apartment has old lead piping, you may have higher than normal levels of lead in your body, particularly if you are using hot water from the tap for cooking or making hot drinks. The reason is that when hot water sits in the pipes it leaches more lead out than cold water does. Replacing old lead piping is expensive, so you can cut down on lead by running water through the tap for a minute or so (to remove the water that has been sitting in the pipes). Better yet, you can drink bottled water.

Lead paint is not nearly as big a danger, since the only way to be contaminated is to eat paint chips or breathe the dust caused by sanding. Also, most lead paint has been painted over in recent years with a nontoxic paint that effectively seals in the lead.

If you live around cats, you may want to get yourself tested for previous toxoplasmosis infection, which can cause birth defects if the infection happens during pregnancy. Despite what you might think, it is actually good if a prenatal test shows that you have already been infected by the *Toxoplasma* bacterium because this means you are immune and can't be infected again. Screening for toxoplasmosis may be advisable at a preconception visit. Most adults have already been infected without knowing it. If you don't know if you have been infected, or a test shows that you have not had toxoplasmosis, then you need to be careful during pregnancy. Don't dig in soil where cats may have defecated, get someone else to empty the litterbox, and always wear gloves and wash your hands after gardening. Also make sure to wash your hands carefully after touching cats (or keep away from them completely).

Likewise, if you have never contracted typical childhood diseases such as measles, German measles, and mumps, or been vaccinated for them, you should get vaccinated now. If you never had chicken pox, you probably do not have immunity because there was no vaccine for chicken pox until recently.

THE BABY LIVES BY SPECIAL RULES

As the baby grows in the womb, his cells are also growing and dividing rapidly, making decisions about the role they will play for the rest of your child's life. Small amounts of unwanted, disruptive chemicals, so small that they are virtually harmless for adults, can be toxic for a developing embryo and fetus. The potential for harm exists all the time but is especially present early in the life of the embryo. Cells that are dividing rapidly are listening for, and responding to, exquisitely subtle chemical signals, which can be overwhelmed by chemicals from outside the womb. Other chemicals can damage the cells' genetic material; since cells are dividing rapidly, effects on a few cells at a critical time in development—the first of my ten principles of programming—can multiply and spread rapidly, affecting many cells in the body.

As we have seen several times before, your developing baby is not just a miniature adult. He is not even a miniature child. One of the principles of prenatal programming is that the fetus lives and grows by his own rules, rules that are appropriate to the development of a vast range of new skills during this exciting time. As an example of the different rules that the fetus lives by, let's look at nicotine, the well-known toxic and addictive product found in tobacco. In the body, acetylcholine is a natural and powerful part of the communication system used by nerve cells. It is one way that some nerve cells talk and issue instructions to their neighbors. The messages that are supposed to pass between cells during fetal life tell cells to grow, divide, or even to die and make way for other cells. Molecules like nicotine can mimic acetylcholine and these disrupt the messages and cause problems. Scientists at Stanford University have discovered that the fetus has its own special nicotine receptor, a kind not found in adults. We don't know what this special receptor does and why it is found only in the fetus, but its presence may mean that nicotine presents a far greater threat and causes far more damage in the fetus than it does in either children or adults.

The dangers of potentially toxic substances might seem one of the scariest topics in this book, because the toxins seem to be all around us, and the damage that they can potentially do seems horrendous. However, take heart in the fact that we are an extremely resilient species. Our bodies have developed powerful methods of coping, of excreting and detoxifying chemicals. As we have discussed, some researchers think that morning sickness may be a way for pregnant women to protect the very vulnerable early embryo, that it is the body's own none-too-subtle attempt to steer you away from harmful substances.

We do not yet know the effects of subtle exposures to toxins in the environment. Much more research is necessary on how the whole baby grows and functions and on how her developing organs interact. For example, the increase in the number of cases of attention deficit disorder in children in recent years may be the result of such subtle damage by exposure to environmental toxins in the womb. Many studies in animals show that the placenta can act as a resourceful barrier to the passage of unwanted chemicals to the fetus. However, the approach for all of us should be not to challenge the system. We need to give the baby the very best, cleanest environment possible.

As with diet, stress, and exercise, a little knowledge can help you learn to avoid the worst risks and to balance the remaining small risks against the practical realities of life. Remember that you are trying to provide the best *possible* first home for your baby, the most suitable womb environment conceivable. You are not trying for the impossible goal of perfection. The good news is that baby-toxic compounds are easy to identify. Avoiding them takes only knowledge and a dedication to providing a toxin-free womb environment. You need to know about:

- Tobacco and alcohol use during pregnancy.
- Setting limits on caffeine.
- Daily household and environmental dangers.

OUR CHILDREN'S FUTURE GOING UP IN SMOKE

Smoking, like drinking alcohol, has been thoroughly discouraged during pregnancy in recent years. But it is still useful to go into the wealth of data on the dangers of tobacco smoke to the fetus, if only for one simple reason: research shows that only one quarter of women who smoke before pregnancy are successfully able to give it up when they become pregnant. Of those who stop smoking, one third will start smoking again before their baby is born. These are astonishing statistics. They soundly crush the protestations of nearly every tobacco company that tobacco smoking is not addictive—it is. Women who are addicted before pregnancy to both alcohol and smoking generally find it harder to stop smoking than drinking. Some researchers have found that the drive to smoke tobacco among regular users is as strong as the desire to consume cocaine among cocaine addicts.

Luckily, most everyone except for tobacco companies knows that smoking is addictive; smokers themselves know this most strongly. We know that smoking is bad for your health and that smoking during pregnancy is terrible. And yet the strength of the addiction is so great that even the most convinced people find it extremely difficult to stop. I've always felt knowledge is powerful and should be your greatest ally in kicking an addiction that you know is harmful. With this in mind, it is useful to remember what research has taught us about the many ways smoking is bad for a developing fetus. The newer research makes it clear that smoking is even worse for your own and your baby's health than previously thought. If you are a smoker, make all efforts to join the growing group of women who are able to stop smoking and stay smoke-free throughout pregnancy. Because of what we have learned about passive smoking, your partner and friends should understand the importance of staying smoke-free as well. It is well worth it and, who knows, perhaps you will have no desire to take it up again after your pregnancy is over. This would confer a double benefit on your child. First, you will help her during the critical stages of development during life before birth. Second, if you stop permanently, your child will be less likely to smoke in later life. There is a strong correlation between parental smoking and children who later take up the habit. By quitting smoking, you will even be participating in a transgenerational passage of the gift of health. Helping your daughter not to smoke when she becomes pregnant will improve your grandchild's very first home.

Tobacco smoke has over four thousand elements, nearly all of which are toxic. We have seen that nicotine, one of the most destructive, is, in fact, a powerful drug, despite what the tobacco companies say. It acts on one of the body's most important neurotransmitter systems, the acetylcholine system. Acetylcholine receptors are critical for activating muscles and controlling the contraction of the heart. We know that smoking raises blood pressure and causes blood vessels to contract. Ultimately, smoking damages the whole cardiovascular system, raising the risk of stroke and heart attack.

Smoking cuts the supply of oxygen to the fetus in three ways. By causing the blood vessels on both sides of the placenta to contract, less blood gets to the place where oxygen and nutrients are exchanged between a mother's blood supply and her baby's blood supply. Also, when a mother breathes in smoke, she is not only breathing in less air, but also bringing carbon monoxide into the blood supply. Carbon monoxide, which is found in high concentrations in tobacco smoke, latches onto the hemoglobin molecule that red blood cells use to carry oxygen through the body, thereby blocking transport of the life-giving gas. Once formed in the red cells, this hemoglobin–carbon monoxide bond breaks down only very slowly, and so the abnormal hemoglobin hangs around for a long time after you smoke. In this way smoking will reduce the capacity of red blood cells to carry oxygen around the mother and to her fetus. Carbon monoxide also comprises 6 percent or more of the exhaust emissions of automobiles. At low speeds, automobiles produce large amounts of toxic exhaust fumes, so the danger is even greater in exhaust-fume-laden cities such as London, Los Angeles, and Tokyo, where traffic speeds may slow down to as low as five miles an hour.

The final and greatest harm is that smoking slows the growth of the placenta. Remember that the placenta is a vitally important organ in itself, not just an anchor for the umbilical cord. Everything good that the fetus receives, everything bad that he must get rid of, passes through or is processed by the placenta. When the placenta cannot do its job adequately, the fetus suffers.

Numerous epidemiological studies vividly illustrate the result. These studies show that women who smoke only a half pack of cigarettes a day have a much higher rate of miscarriage. The studies also reveal that smoking during pregnancy results in an average reduction in birth weight of around 150 to 320 grams (0.33 to 0.7 pounds). This reduction may seem small, but you need to compare it to the 150-gram average birth weight reduction in babies who experienced the Dutch Hunger Winter in the womb. It seems that for the subjects in this study, a smoking mother is as bad or worse than a starving mother in terms of growth in the womb.

Statistics show that problems associated with smoking during pregnancy are widespread. In 1990, the US surgeon general reported that smoking was the major cause of 26 percent of all low–birth weight children born in the United States. Maternal smoking was also the cause of 14 percent of premature deliveries and 10 percent of fetal and infant deaths.

Tobacco smoke also disrupts fetal sleep, particularly the dream state known as rapid eye movement, or REM, sleep. Nicotine at the concentrations provided by smoking will halve the amount of REM sleep in newborn rats. We don't know exactly what REM sleep does in the developing brain, but there are many clear indicators that show it is important for brain development. About 80 percent of the sleep time of the third-trimester fetus is devoted to REM sleep, as opposed to about 20 percent of our sleep time as adults. Disrupting REM sleep in young rats has been shown to affect their intelligence as adults. One theory is that REM sleep is the brain's way of testing out neural circuits as they form, sorting out the good from the bad, so that the neural microprocessors in the brain can function correctly after birth. If so, disrupting REM sleep during brain development is one of the worst things that can happen to your baby during the third trimester.

The negative effects of being exposed to smoking in the womb can persist into childhood. The children of mothers who smoked during pregnancy have a higher incidence of respiratory diseases and other minor ailments. Nicotine disrupts sleep after birth as well. There is a strong correlation between smoking during pregnancy and "crib death," or sudden infant death syndrome (SIDS). Children with attention-deficit hyperactivity disorder (ADHD) are much more likely to be the children of mothers who smoked while pregnant than not.

One reassuring note is that even though there are the negative effects of tobacco throughout pregnancy, they seem to be greatest during the last trimester. So even stopping late in pregnancy can help your baby. If you are smoking and already pregnant, don't fatalistically think that you have already done the damage so you may as well keep smoking. The evidence is very clear: stopping smoking at any time during pregnancy will have a strong positive effect.

SECONDHAND SMOKE

You don't have to be smoking yourself to expose the fetus to dangerous tobacco smoke. Every year it is becoming clearer that breathing secondhand smoke can be just as dangerous as actually smoking a cigarette or cigar. The fact is that only 15 percent of tobacco smoke goes into the smoker's body. The other 85 percent enters the air around the smoker. Some of this comes from smoke the smoker exhales after a puff. Even more dangerous is that portion of the smoke that comes directly from the burning tobacco, without being cut by a cigarette filter or the smoker's lungs. In an enclosed room, the concentration of smoke can become so great that breathing the air is as bad as smoking. And just as with smoking itself, the fetus is being exposed to tobacco smoke.

To cut your own (and your baby's) exposure to secondhand smoke, you must be in a smoke-free environment. Simply opening a window in a room, or using one of those smoke-filtering devices, won't make a substantial difference in the level of smoke in the room. Ask for nonsmoking tables at restaurants and a smoke-free environment at work (luckily, most workplaces are smoke-free now). And since you spend most of your time at home, you need to ask people to not smoke in your house. If a spouse or housemates smoke, they should smoke only outside the house. Better yet, push them to take this opportunity to quit altogether. Since secondhand smoke continues to be a problem for infants after the pregnancy is over, now is as good a time as any to get everyone in the household to give up smoking.

QUITTING SMOKING

Quitting is difficult, but not impossible. There are many formal programs to help people quit, but experts say most use various mixtures of the three methods that really work. Those are:

- Using a nicotine patch or nicotine gum.
- Getting support and encouragement from those around you.
- Learning how to handle both the urge to smoke and stress in general.

NICOTINE PATCH OR GUM

Since nicotine is itself a major part of the problems caused by smoking during pregnancy, using a nicotine patch while you are pregnant is also out of the question. Again, preparing for pregnancy is the right way to go. Nicotine patches or gum can be used as a useful step to quitting *before* pregnancy begins. The nicotine patch and nicotine gum are powerful allies in the struggle to quit, providing one more good reason that you should be trying to quit long before you get pregnant.

SOCIAL SUPPORT

Enlist people to quit with you, especially if they live in the same household. If your husband or partner also smokes, try to persuade him to quit too. It's one very small but important way that he can share the burden of pregnancy, and he will not be exposing you and the baby to secondhand smoke. Social support also means lining up sympathetic friends to talk to when you get a craving to smoke, people who will cheer you on when you succeed in cutting back or eliminating tobacco use day after day.

CONTROLLING URGES AND STRESS

The urge to smoke is a web of biological addiction and psychological need. People smoke most when they are under pressure, when they are stressed, when they feel sad or low. A large part of quitting smoking is learning to control the feelings that increase your psychological drive to smoke. That means using the stress-reduction techniques I described in chapter 4, and any other techniques that work for you.

Avoid situations and places that remind you of smoking. Being in those situations will only make you think of smoking more strongly and increase your urge to smoke. Also, stop smoking first in the places you have to be most often, like at work or in your car.

Of all the changes that come your way as a result of pregnancy, quitting smoking can be one of the hardest. I cannot emphasize enough that it is very important to your success to try to quit long before pregnancy begins, rather then when all the other stresses of pregnancy are upon you.

There are a number of resources you can turn to to get professional help in quitting.

ALCOHOL: JUST ANOTHER LITTLE DRINK?

Alcohol is another powerful drug that has extensive effects on your vascular system and your body's metabolism, in addition to the obvious effects it has on many functions of your brain, such as thinking and fine muscle movements. The presence of the surgeon general's warning on each bottle of alcohol makes it hard to avoid the knowledge that alcohol is harmful during pregnancy. But there also been some discussion about the healthful effects of red wine and other alcoholic drinks, and some suggestions that a little wine or alcohol is not as harmful as we have been led to believe. There have even been some suggestions that small amounts of alcohol might do little harm late in pregnancy and do a lot of good in relieving stress, which we have seen can be harmful in itself. So what is the real story?

First of all, we know that alcohol in moderate to large amounts is harmful during pregnancy. Alcohol cuts the blood supply to the fetal brain and affects how nerve cells communicate with each other. Cells therefore get bad information during development. As a result, those same cells make bad decisions about when to be active and when to rest.

The harmful effects of alcohol on many aspects of fetal development have been known for many years. In 1968, scientists from France were the first to describe fetal alcohol syndrome (FAS), in which babies repeatedly exposed to alcohol in the womb are born with mental retardation and physical abnormalities. FAS babies can often be recognized right after birth and are easily identified within a few months of birth. They have slow growth during fetal and postnatal life and lack the ability to catch up in growth, even when provided with excellent nutrition after birth. This inability to catch up is likely due to a decrease in the number of cells in the baby's body, due to the alcohol's inhibition of cell division early in development. This is why early exposure to high levels of alcohol is particularly harmful. In addition to their small, thin bodies, FAS babies have small heads and brains, accompanied by mental retardation, behavioral irritability, hyperactivity, bad coordination, and learning problems. The faces of FAS babies usually have small, widely spaced eyes and an upturned nose with a flat bridge.

Now, most of us born before the mid-1980s had mothers who drank small amounts of alcohol during pregnancy, and none of us suffer from FAS. Clearly, alcohol very occasionally is not going to give your baby FAS. In recent years there has been a somewhat justifiable backlash against the implied idea that mothers who even touch a few drops of alcohol are somehow unfit. Some doctors have even said that since the biggest problems with alcohol are early in pregnancy, perhaps the benefit of reduced stress from a small drink late in pregnancy outweighs the harm of the alcohol. The fundamental question is, how much alcohol is too much?

What we know is that heavy drinking (more than three drinks a day) and binge drinking (five or more drinks on any single day) can definitely produce symptoms of FAS or a milder form, called fetal alcohol effects (FAE). That means that this level of drinking challenges developing cells sufficiently to produce obvious damage. At this point, we don't know what the effect of light drinking, say one drink per day, may be on the fetus. The Institute of Medicine reiterates this position in their official report on alcohol and pregnancy: "At present, there is uncertainty whether minimal alcohol intake during pregnancy could be associated with any degree of injury to the baby." One study does suggest that the likelihood of miscarriage in the first trimester is doubled in women who drink as little as one ounce of alcohol twice a week. It is also worthwhile to remember the fundamental idea of prenatal programming: any challenge to developing cells will inhibit them in their natural course and will alter their behavior, often permanently. Alcohol certainly presents a challenge to growing cells. If a mother is also sick, or is exposed to tobacco smoke (via smoking or secondhand smoke), or lives in some other situation that reduces the supply of oxygen to the baby, the effects of alcohol will be that much more potent.

After considering all the evidence, I would say that there is probably some minimal amount of alcohol that is not detrimental to the growing fetus, but we don't know what that level is. Perhaps one drink a day is okay, or perhaps the safe level is only a sip a week. Some studies suggest that as little as two drinks a week may result in increased agitation and stressful behavior in newborn babies. We just don't know, so the best advice to take is probably: "In ignorance, abstain." Whatever benefits there are in a little alcohol, such as stress reduction in late pregnancy, can be provided by the nonalcoholic stress reduction techniques I described in chapter 4. If you are practicing these techniques throughout pregnancy, you shouldn't need alcohol to relax and cut stress in the third trimester.

CAFFEINE

Coffee. Tea. Colas. Much of the world's population relies on the stimulating effects of caffeine to make it through the day. For many people, these drinks are one of their great pleasures, in addition to being a great source of craving. During pregnancy, women usually feel they need caffeine even more. The natural hormones of pregnancy set the body in a slower, energy-storage mode, just like the effect you get after eating a big meal—the "postprandial stupor" that helps encourage the storage of calories you have just consumed. So caffeine seems doubly necessary, like amphetamines for the narcoleptic, to get you through the day. It is of interest that despite the craving for caffeine during pregnancy women may find that they consume smaller amounts of beverages containing caffeine. There is a good explanation for that change. As part of the adaptation to pregnancy the liver and kidneys change the way they handle caffeine, as a result half the dose of caffeine produces the same blood concentration during pregnancy as was produced before pregnancy. This difference between the way pregnant and nonpregnant women handle caffeine has been demonstrated in many studies. Knowledge of increased sensititvity to drugs in relation to caffeine helps to remind us that the way the body handles a wide range of drugs can change substantially during pregnancy. It may be necessary to adjust any medications across the course of pregnancy. Always get professional advice from your care giver. But, no doubt about it, caffeine is a powerful drug and we must always be mindful of its potential effects on the developing embryo and fetus.

Since caffeine is so widely consumed, there have been a fair number of studies of its effects on pregnant women. Many of these studies have been contradictory, but after sorting out all the facts, it seems that moderate caffeine consumption is not harmful to the fetus.

First, a word about how caffeine works. One of the ways your body regulates the activity of neurons is through the presence of a chemical called adenosine. Like the ash that comes from burning a log, adenosine is one of the chemicals left over after cells burn the body's primary energy molecule, adenosine triphosphate. One theory is that very active neurons, such as might occur during an emotionally wrenching day, consume a lot of energy, leaving a larger than usual amount of adenosine. Adenosine then puts the brakes on neural activity, leaving you tired and lethargic. Caffeine blocks the action of adenosine, taking the brakes off cell activity.

So caffeine speeds up cell activity. That's why we feel less tired, more energetic, even more optimistic. One study showed that caffeine can even act as a simple antidepressant. Many people consider caffeine to be a lifesaver. The problem is that the body becomes tolerant of caffeine when exposed to it regularly. You require more coffee to get the same stimulating effect. One cup a day can quickly turn into a four-, five-, or six-cup-a-day habit for some people.

The first clinical research studies that raised a warning signal about caffeine use in pregnancy seemed to show a subtle correlation between miscarriage and women who used caffeine during pregnancy. This was very troublesome for many women, and for some time physicians recommended against having any coffee for the duration of pregnancy. But other studies confused the issue when they did not find any correlation between caffeine and miscarriage. New ideas on the relationship between the ability to tolerate coffee and miscarriage suggest that many women feel so nauseated at the smell and taste of coffee that they cannot stand the stuff, even though they imbibed regularly before becoming pregnant. This subgroup that stops drinking coffee may already have fewer problems (unrelated to caffeine consumption) than those who can continue to drink coffee, and this makes it very difficult to decide whether coffee has any direct overall harmful effects.

A recent study at Johns Hopkins University on the frequency of miscarriage tried to get around this problem by studying not only whether women drank coffee or not, but also how many cups they drank. The study found no increased risk of miscarriage in women who consumed one to three cups of coffee a day, but there was an increase in miscarriage risk among women who consumed five or more cups a day.

The overall results of these studies seem to show that caffeine is safe in terms of miscarriage risk if you limit yourself to one or two cups of coffee a day. But having a lot of coffee is questionable. In addition to the potential increase in the risk of miscarriage, a lot of coffee will interfere with your baby's sleep patterns in the womb. It's just as bad or worse to deprive your child of needed sleep before birth as it would be to give him coffee and deprive him of sleep after birth.

If you are having a lot of coffee or tea (more than five cups a day), you may have trouble cutting down. Your body has become habituated to caffeine, and cutting back may give you headaches and make you irritable and tired. Just remember, though, that the baby is also becoming habituated to caffeine, and that he will go through the same uncomfortable decaffeinating process when he makes the journey from the womb to the outside world. This is not an easy transition to make in the best of circumstances, and it is a good idea to spare your baby the burden of breaking a coffee addiction after birth, in addition to all the other pronounced lifestyle changes he will face at that time.

Coffee provides the most concentrated form of caffeine, but tea, soda, and even chocolate contain a fair amount of caffeine too. Coffee usually has about 90 to 150 mg of caffeine in an eight-ounce cup, whereas tea has about 60 mg. A standard twelve-ounce can of cola usually contains 30 to 50 mg of caffeine. One ounce of chocolate contains about 25 mg of caffeine, which means you would have to eat an awful lot of chocolate to get the equivalent of five cups of coffee.

Heavy caffeine users often experience headaches, irritability, fatigue, and even a sick feeling when they try to stop having their daily dose of coffee, tea, or soda. Cutting down can be easier if you

- Use your one- or two-cup allotment when you need it most, perhaps in the morning.

- Drink lots of water during the day.

- Drink decaffeinated coffee, which has a slight amount of caffeine but not enough to be unhealthy in pregnancy. Recent studies have shown that many of the harmful factors other than caffeine in coffee are decreased by filtering. In addition, although it may not please the gourmet in you, instant coffee is probably safer than strong, unfiltered coffee.

- Drink decaffeinated tea. You should check the ingredients of herbal teas carefully, for the reasons I have listed in the section below on health foods.

- Start cutting down on caffeine consumption before pregnancy begins.

TOXIC CHEMICALS IN DAILY LIFE

We are surrounded by chemicals that are toxic or carcinogenic in sufficient quantities, but seem to be fairly innocuous in small amounts. It is a cause for concern that many food additives, pesticides and other chemicals have never been throughly tested for their effects on reproductive function and especially on the developing fetus and embryo. Studies that have been conducted have usually used adults as test models. For an embryo or fetus, the risks of these same chemicals are much greater. Remember, your baby lives and develops in the womb by a set of biological rules that are different from yours and mine in adult life. We don't yet know everything about those rules, and we certainly don't know all we need to know about exposures that alter the trajectory and sequence of developmental events.

Let's use gasoline as an example. Benzene and other major components of automobile gasoline have the ability to alter DNA and are therefore mutagens and/or carcinogens. DNA mutations in adults are mostly innocuous or easily repaired by the body—even if you are an oil field worker, the chances are that any given mutation in a single cell won't give you cancer. For an embryo, though, a single cell with one mutation can divide into many thousands of cells which then carry the same mutation. The process of division can even spread the error and cause more mutations. The risk of cancer or some other problem therefore increases. So it pays to be extra careful when you are pregnant, and to try harder to avoid problem chemicals. The chances of harm are still slight, but also slightly increased.

This means that you should try to not pump your own gas while you are pregnant, because the fumes present a small risk. You should also not be handling paint thinner or other organic solvents. If this is part of your job, you should ask to have your duties changed. All the usual cautions on labels apply, only more so. Don't put on nail polish or remove it in an enclosed space without ventilation. If you find someone next to you on an airplane painting their nails, you can let them or the flight attendant know that there are Federal Aviation Administration regulations forbidding this. Try not to use oil-based paints, and if you do, make sure there is plenty of ventilation. Anytime such chemicals make you feel even slightly dizzy, get away from them. This also

Prenatal Prescription   22.50
0060197633
DISCOUNT   25.00 - 2.50
Magazine   5.35
071896469622
DISCOUNT   5.95 - .60

SUB TOTAL   27.85
SALES TAX   1.56
TOTAL   29.41
AMOUNT TENDERED
AMEX   29.41
CARD #:
AMOUNT   27.85
AUTH CODE   045227

READERS' ADVANTAGE SAVINGS   3.10

holds true for household cleaners like bleach or ammonia. When the nesting instinct kicks in, you may find yourself with a strong desire to sterilize your living space, but try to use ordinary soap, or get someone else to do the really heavy housecleaning.

There are also dangers in the kitchen that are deemed a slight risk for adults, but are a greater hazard for a fetus. Smoking foods or charring them on the barbecue creates carcinogenic compounds, so you should avoid them. Check labels for the presence of the preservatives sodium nitrite or sodium nitrate. These compounds are changed in the body to compounds called nitrosamines, which look enough like the basic building blocks of DNA that they can induce genetic errors. When exposed to the sun, potatoes create toxic compounds in their skin and eyes to protect themselves from insects, so you should never eat potatoes that have a greenish color (a really green potato can carry enough poisonous compounds to make even an adult sick).

There has been a lot of worry and discussion of the dangers of pesticides, but most of the toxic effects have been found in children that live around farming communities or in the adults who handle the sprays. The minuscule amounts that are left on vegetables shouldn't be a problem if you wash your vegetables with soap and water (some pesticides are applied in oil-based solutions that won't wash off with water alone), or you may just decide that you will worry less if you switch to organic foods during pregnancy. In any case, don't stop eating fresh vegetables. They have been shown to contain anticancer compounds that protect against genetic mutation.

By all means, don't panic and think that everything around you is an imminent threat to your child's life. Just take more of the sensible precautions than you normally would, be especially vigilant about cautions you read on warning labels, and don't panic yourself about things that are beyond your control. Remember, you are trying to create the optimal womb environment, not the perfect womb.

DRUGS PRESCRIBED BY PHYSICIANS

Few discussions are less well informed than those about the unwanted effects of chemicals on our bodies. A major part of the problem is the difficulty of determining which compounds are "normal biological materials" and which are not. The tobacco industry has long claimed that nicotine is *not a drug* because it is a normal component of our bodies. The inference is that therefore it cannot harm us, or cause addiction. We should remember that all disease springs from having too much, or too little, of normal compounds. Unscrupulous purveyors of snake oil and other quack remedies often make a similar claim, saying, "It's just a normal compound in the body . . . it's the natural remedy." Just because it's natural doesn't make it good.

Let's first look at those compounds that your physician may need to prescribe for you during pregnancy. All FDA-approved medications are specifically synthesized, purified, designed, and manufactured in order to have powerful biological effects. For this reason, pregnant women and physicians are usually careful when considering whether to use a certain drug or medication. Testing of the drug has included screening for the ability of the drug to produce birth defects. Despite all these precautions, we are now learning about how individual reactions to drugs can differ greatly as a result of each person's genetic composition. In pregnancy, we have to remember that we are considering two individuals, and the interactions among the two of them and the drug.

We can never be certain how an individual fetus will respond to even approved drugs, so physicians only prescribe medications when there is some genuine reason related to your health that must be addressed. During pregnancy, physicians must weigh the costs and benefits of drug treatments even more carefully than they do before your pregnancy. When they decide you need to take pharmaceuticals, there is usually a strong reason for their use. Anti-epileptic drugs can be life-saving for a pregnant woman. Drugs to lower blood pressure are sometimes essential for women who have very high blood pressure late in pregnancy.

Despite these benefits, both you and your doctor have to pay attention to the deficit side of the equation. Effective drugs are effective precisely because they are biologically powerful and usually last longer in the body than the natural compounds that they mimic. Most prescription drugs have some side effects, which also have to be taken into account. Prescription drugs are always well tested in animals in order to avoid these potentially harmful side effects. My advice is to avoid medications in general if possible, but if your physician feels they are necessary and you work with him or her to identify risks and benefits, you can take the medications with minimal risk of harm.

Whether or not to use antidepressants can be a particularly tough decision for you and your doctor. Depression is a serious, life-threatening illness. More than just "the blues," depression can undercut your sense of self so severely that suicide seems to be the only option. Even if a depressed person does not feel suicidal, she may feel hopeless and stressed, which we know can cause its own problems for the developing fetus. Obviously, these risks have to be taken into account when considering whether antidepressants should be continued during pregnancy for someone suffering an existing depression, or begun when depression first strikes during pregnancy.

One problem is that we don't know all the risks of taking antidepressants during pregnancy. Luckily, a fair amount of research has shown that there are many existing antidepressants that do not cause malformations in the developing fetus. They are not directly toxic to cells when taken in prescribed amounts and when other precautions are observed. What we don't know well enough is how these medications may affect the developing brain. The developing brain wires itself up partly according to the nature and amount of activity going on in the developing nerves themselves. Since antidepressants affect nerve activity, we can assume, although we don't know, that antidepressants change how the fetal brain is wired up. Some newborns exposed to antidepressants before birth show signs of withdrawal shortly after birth. Animal experiments have suggested that nerve activity does change for at least a week after birth when the animals are exposed to antidepressants in utero. We don't know if the same changes occur in humans, or if these changes are permanent. If you are afflicted with depression, finding a course between these risks is something you will need to evaluate with your psychiatrist and obstetrician.

OVER-THE-COUNTER DRUGS

By regulation, drugs that can be sold over the counter are for the most part considered to be safer than those that require a prescription. However, there are some common household medications that can cause problems for the developing fetus.

The most important consideration is that you must be very careful to follow recommended dosages. Just because these medications are fairly low risk at the correct dosages doesn't mean that the same medication will be harmless at higher dosages. The painkiller acetominophen is a good example. You may, like millions of others, regularly take acetominophen for headaches and minor aches and pains. But you may not know that an overdose of this over-the-counter painkiller is highly toxic to the liver and can be deadly. Furthermore, you may not know that alcohol keeps the body from clearing this toxin—one study found that people having as little as three times the recommended dosage of acetominophen and a few drinks show signs of liver damage. Antihistamines in high dosages can also cause problems with anxiety and jitters.

You should avoid some medications even at recommended dosages. Aspirin is one of these. Aspirin can interfere with blood clotting, which is why people take it to lower their risk of coronary artery blockage. This same anti-clotting property can cause small hemorrhages, especially in the last trimester of pregnancy and during or after birth. In addition, aspirin interferes with certain metabolic processes. Caution with some of the new anti-inflammatory drugs is also warranted: they are powerful because they work on the same enzymes that play an important role in the process of labor and help to keep special parts of your baby's circulation open.

The cardinal rule for taking medications is that you should check with your doctor before you take any medication, whether prescription or over-the-counter. If you need painkillers, or indeed any other drug during pregnancy, get your physician's recommendation before you start. If you have seasonal allergies and will enter allergy season during your pregnancy, ask your doctor to recommend an allergy medication before you need it, so that you won't feel compelled to grab an over-the-counter allergy medication that you know nothing about because you can't get an appointment with your doctor soon enough.

HEALTH FOODS AND NUTRITIONAL SUPPLEMENTS

Throughout this book I have told you about the benefits of eating healthy, unprocessed foods and staying away from toxins during your pregnancy. The idea that "natural" things are good for the baby might lead you to health food stores for "natural" treatments as an alternative to conventional medicines. But the fact is that these health foods can be very unhealthy for a fetus, and even for adults. While many health foods and nutritional supplements can be good for you, the idea that their "naturalness" makes them automatically safer and healthier than manufactured drugs is false. Many natural herbs and fungi have powerful biological effects and have been used for thousands of years as drugs. Indeed it is their very power that has made them attractive over the years. I am often amazed that the very people who are so skeptical of the claims of the pharmaceutical industry are so blindly trusting of the health food industry.

Just one example of a common plant that harbors a powerful drug is the foxglove plant, which contains digoxin, a drug that is still used to treat failing hearts. Other plants have been used by native healers to induce abortions. One powerful plant that I have studied is the corn lily (*Veratrum californicum*), which is sometimes eaten by pregnant sheep. It contains a potent neurotoxin that zeros in on and destroys the part of the fetal brain critical for initiating labor. When pregnant sheep eat this plant, the labor process is blocked. Some studies show that the pregnancy can be greatly extended, to a period equivalent to a fifteen-month pregnancy in women. I hope these plants would never be found in nutritional supplements, but they illustrate how powerful and even dangerous everyday plants can be.

The second reason for concern about health foods is that they are relatively unregulated. Despite the fact that natural substances can contain powerful drugs, the Food and Drug Administration still views them as foods, not as drugs. So while manufactured pharmaceuticals have to go through many years of testing in cell systems on the laboratory bench, in animals, and in human clinical trials, there is no testing at all for naturally grown substances. In both Europe and America you can buy pills with bacteria that are supposed to colonize your gut and help with digestion. Not only has there been no testing of this claim, there is no quality control to ensure that the bacteria are what they are supposed to be. The few tests that have been done have found that these pills often contain bacteria resistant to antibiotics, so that if harmful bacteria were to slip into these pills and make you sick, doctors might not be able to stop an infection before it kills you. Regulatory agencies' benign view of the situation is beginning to shift due to a few recent incidences in which some people have become sick or died after consuming nutritional supplements. One recent example is a substance called GBL (gamma butyrolactone), a compound that has been promoted as a sleep aid for people who have been working out and are sore. The FDA tracked almost 150 cases in which people had seizures after taking GBL, and one death. The compound got the most notoriety when the forward for professional basketball's Phoenix Suns, Tom Gugliotta, almost died after taking a "health food" that contained GBL. Gugliotta had no idea he was taking GBL, because it was just one ingredient in a product sold as a natural aid for sore athletes.

Which brings up a third concern with such supplements: supplements and normal foodstuffs like vitamins often interact with each other in unforeseen ways. Health food stores can sell any natural substances, in any combination. If you did happen to know exactly how everything in the health food store affects you,

you would not necessarily know whether the stuff you were buying contained the substances you want to avoid.

This is why I recommend against drinking herbal teas during pregnancy. While herbal teas have a very benign image, they often contain a witch's brew of various herbs. It's hard to know exactly what you are drinking. Tea brewed from the zoapatle leaf is used in Mexico to induce abortions. While knowledgeable people would never put this leaf into any herbal tea, it's possible that someone might unknowingly put it into their herbal tea mix because it creates an interesting scent or flavor.

Caution with herbal teas also carries over after pregnancy, during breastfeeding. Even teas that are advertised as being beneficial for breastfeeding can contain potentially harmful substances. One such tea contains comfrey leaf, even though parts of this plant can cause liver damage. This and other teas also contain fennel, which is a weak diuretic, and fenugreek—an herb that is claimed to increase milk production but also is known to lower blood sugar dramatically.

For safe herbal teas, stick to those that contain mostly mint, ginger, rose hips, orange spice, or primrose. If you don't know anything about the ingredients or the ingredients are not mentioned, it is best to stay away from that tea.

DRUGS OF ABUSE

After the issues we have considered in this chapter, I hope it is not really necessary to warn anyone, especially pregnant women, about the dangers of drugs of abuse such as cocaine and marijuana. Cocaine is a powerful drug that cuts blood supply to the fetus and interferes with communication between cells in the developing brain. Marijuana smoke has many of the same toxic compounds as tobacco smoke, in addition to psychoactive chemicals that interfere with natural brain development. In fact, since most drugs of abuse target nerve cells and change the way we think, they have the potential to make permanent changes in thinking ability when a fetus is exposed to them in the womb. Any woman who is taking any of these "recreational" drugs should halt their use immediately—and should really stop using them before becoming pregnant if possible. Get help from a support group or a professional if you can't stop on your own, but stop.

BEST ENVIRONMENT WITHIN AND WITHOUT

For most people, becoming knowledgeable about environmental chemicals that carry an extra threat for your baby is no cause for panic. As I have said, the risks are very manageable. You don't need to radically alter your environment to protect the child inside you. All you need to do is make an extra effort to take care of yourself, to do the things you should be doing and observe the cautions you should be observing anyway. With a little knowledge and a little judgment, you will be able to provide your child with a womb in which the risk of injury from environmental toxins is negligibly small.

7

Preventing Premature Delivery

Your baby's sojourn in your womb is a miraculous and tumultuous period of change and growth, during which he will develop the vital skills that prepare his body for matriculation into the harsher and less protected world outside. Bathed and buoyed by warm, salty fluid, his body begins to mature, his organs start to work on their own and integrate with each other to achieve that marvelous goal of independence. Every part of this preparation is carefully choreographed, so that each event happens in a certain order. Each element of the baby's life-support system is developing at a different pace, but they are all scheduled to be on-line and working at the end of forty weeks, ready for birth and separation from the placenta and for the beginning of a new relationship with you, his mother.

In your womb you have created the perfectly adapted environment for these preparations. One of the gravest threats to a child's well-being and future health is leaving this perfect environment before the scheduled preparations for birth are complete. Babies born too early may not yet be fully ready for the complex challenges posed by an independent life. Such premature babies have not yet fully developed their lungs, kidneys, and other organs. If the fetus has inadequate time in the uterus to prepare thoroughly for the great adventure of life, there can be long-term consequences for his health. Since the brain is growing and organizing itself rapidly during the second half of pregnancy, many of the problems that arise from prematurity are neural. Dr. Maria Fitzgerald at London University has been studying the development of various nerve fibers before and after birth. She has shown that it is very easy to overstimulate premature babies, probably because the immature nervous system has a preponderance of nerves that produce an irritable response.

There are different ways to define prematurity, but generally we say that babies born before thirty-seven weeks of gestation are considered premature. Babies born one or two weeks before the due date (or a week or so after) are close enough that there is not significant risk to their health. Premature births represent only 10 percent of all births in the United States, but they account for a striking 50 percent of all long-term handicaps. That means that in the United States there are 400,000 children every year who are put at risk by coming out of their maternal home too early. Unfortunately the incidence of prematurity in the United States is increasing and represents one of the most significant health disparities thoughout society. While all other causes of infant mortality fell over the decade 1988 to 1998 death as a direct result of preterm birth or being of low birth weight actually increased. These prematurely delivered children account for the majority of childhood health problems and long-term handicaps. It is therefore very important that children be given every chance to make the right decision about when they should be born.

The exact consequences of being born too early depend on

just how prematurely a baby is born. For babies born only a few weeks early, the long-term effects are usually slight. For babies at the other end of the scale, those born a little over halfway through gestation and weighing one pound or less, the threats are very real both in the short term and in the long term. All prematurity has some health consequences, because the baby's body is forced to grow and mature in a world for which he wasn't fully ready. Remember, one of the prime rules of programming is that when cells are challenged early in development, they make adaptations that affect how well they function in the long term. However, the good news is that we are an extremely resilient species. Isaac Newton, the great physicist, was born prematurely and apparently small enough to squeeze into a pint pot. Winston Churchill is another great mind who was supposedly born prematurely.

Premature labor occurs when contractions start too early. Through much of a normal pregnancy, women experience an occasional, gentle tensing of the muscle in the wall of the womb. These muscle tensings are a normal part of the womb's preparations for labor and are called contractures or Braxton-Hicks contractions. For some reason, in premature labor these contractures grow stronger and transition into full-blown contractions before the full forty-week gestation period is over. Once real contractions start, it is very difficult to stop them before labor is completed.

We are still in the early stages of understanding both the mechanisms that bring about this switch from contractures to labor contractions and how we can stop them from happening, but there is new information that women should know in order to monitor their own risk of delivering early. In a few specific situations, there are some steps women can take to lower their risk of premature labor and delivery:

- Fight bacterial infection in the body.
- Protect your baby from physical stress.
- Fight chronic psychological stress in yourself.
- If you smoke, stop.

After researching this area for over thirty years, I am sure that premature labor doesn't have one cause, just as a high body temperature can be the result of the flu, a bacterial infection, or a strenuous workout. Premature labor may be the result of stress, infection, genetic predisposition toward prematurity, or other as-yet-unknown causes.

DECIDING WHEN TO BE BORN

Let the child decide when to be born? It's true—most people don't understand that it is the fetus and not the mother who decides when it is time for birth. The child in the womb uses cues from the environment and from his developing organs to decide when it's the right time to come out. That's why it is so important to make sure that the womb is the friendliest, most nurturing environment possible, so that the fetus will stay there comfortably as long as possible. A wide variety of animal studies indicate that as the baby matures in the womb, each critical organ system sends chemical and nerve signals to the brain informing the brain that all the organs in the baby's body are ready . . . all signals are go. These signals all come together in the hypothalamus, a very ancient part of the brain that controls the body's critical functions like hunger, sleep, heart rate, and so on. You may remember from chapter 4 that the hypothalamus is also an essential part of regulating the stress response. When all signals are go, this master control center sends out mail packets of hormonal chemical signals. These little chemical snowballs grow to become a biological avalanche—the unstoppable hormonal and muscular activity of labor.

I have spent many years studying, these fascinating interactions between mother and baby in sheep and monkeys. As a result of these studies, we know that in sheep the go signal from the baby's hypothalamus spurs his adrenal glands to release the stress hormone cortisol. This hormonal signal from the baby causes the placenta to switch from producing progesterone, a hormone that sustains pregnancy, to estrogen. Estrogen then stimulates many of the systems that the mother will use to start labor. So birth is a cooperation between baby, placenta, and mum. Throughout pregnancy, progesterone has quieted the muscle wall of the womb, thus helping to continue the pregnancy. Estrogen, in contrast, stimulates the muscles of the womb to contract. For normal labor and delivery to occur, major changes must occur in the cervix—the collar of tissue that acts as the baby's gate to the birth canal and the outside world. Estrogen also choreographs the forces that make the cervix soften and dilate and that cause the membranes around the baby to rupture. Primates such as humans and monkeys use a similar but slightly different system. In the fetal primate, the adrenal glands release a similar steroid, androgen, which the placenta turns into estrogen. So in sheep, humans and other animals, the result is the same: the level of estrogen slowly rises, overcoming the muscle-quieting effects of progesterone. Estrogen stimulates the mother's brain to produce the hormone oxytocin. Oxytocin is the most powerful stimulus to contractions of the womb that has been discovered. Oxytocin causes the muscles in the womb to switch from the Braxton-Hicks contractures that have gently squeezed the baby periodically throughout late pregnancy to the strong, full-blown contractions of labor.

At the same time, the baby's adrenal steroid hormone cortisol is instructing the fetal lungs, liver, gut, brain, and kidneys to make the last, detailed preparations for an independent existence in the outside world. Obstetricians sometimes make use of this knowledge when they diagnose that premature labor is imminent. If they think a baby will be born prematurely, they give the mother a course of injections of synthetic steroids that cross the placenta and speed up maturation of the baby's vital organs—especially his lungs, without which he will not be able to get enough oxygen. The steroids work on the developing lungs in as little as forty-eight hours.

While the baby decides precisely which day he is ready to be born, the mother determines the time of day. At the end of pregnancy, the baby is sending out increasing signals over a period of days. At the same time, the rising levels of estrogen makes the mother's uterine muscles more sensitive to oxytocin from the mother's brain. In other words, the baby comes knocking at the door, and the mother opens it.

It's no news to either obstetricians or women who have had babies already that the process of birth most often begins in the evening. Women often report that one or two evenings before labor actually began, their contractions increased in strength and frequency, changing to true contractions for a time, only to fall off again a few hours later. This is because although Mom has registered her baby's signal that he is ready to be born, her response is not yet well enough coordinated and strong enough to initiate labor. She opens the biochemical window to labor, but not quite enough. Later the same night, the window closes.

The repeated nighttime switch from contractures to labor contractions has considerable value. First, it is a way of synchronizing the fetus to the outside world's schedule a few days before birth. All through pregnancy, the baby has been getting some signals about the twenty-four-hour day in the world outside his cozy home, but the evening contractions are a final reminder that reinforces the message. Good synchronization prepares the baby for his rhythms of feeding and sleeping and wakefulness, which must be integrated with his mom after birth. The nightly contractions also help the fetus become gradually prepared for birth over several nights before labor actually begins. For one thing, the gradually increasing pressure pushes the baby down into your pelvis, gently molding his head to the shape of the birth canal, something that has to take place before birth can occur. The repeated pressure on the cervix also helps to soften and dilate this important exit from the womb before labor begins.

Once labor begins in earnest, there is no turning back. Whereas most reactions in biology, like the stress reaction, have negative feedback (dampen themselves down) so that the reaction doesn't run away with itself, the chemistry of labor is self-reinforcing so that each contraction sets off processes that increase the contractions that follow until the baby and placenta are expelled.

WHY ARE SOME BABIES BORN TOO EARLY?

We don't know all the reasons why some babies are born prematurely, but we do know that premature birth often happens because the baby is under some sort of stress. It may be that when conditions in the womb environment become overly threatening to the baby's health, the child has a better chance of surviving outside the womb rather than within it. The most common reasons for this are infection in the womb or birth canal, and stressful situations for either mother or child.

Infection in the vagina or uterine cavity is believed to play a part in one-third to one-half of the cases of prematurity. If we look at only the most serious cases of prematurity—babies born before the thirty-second week of pregnancy, infection is found to play a role in more than 40 percent of all cases. An abnormal number of bacteria in the vagina and reproductive tract is called bacterial vaginosis. Sometimes there are even bacteria in the amniotic fluid. Bacterial vaginosis may cause premature labor because the fetus interprets the infection as a threat and decides to end the pregnancy sooner than scheduled, or because the chemical changes associated with infection might short-circuit the controls that inhibit labor. Bacterial vaginosis doesn't always lead to premature labor by itself—about 20 percent of women have it sometime during pregnancy, and most of these women do not go into labor prematurely. But it can be part of the process that jump-starts labor and significantly raises the risk of prematurity.

Stress in either the fetus or the mother may also lead to premature labor. When the baby finds that there is no more room to grow or when there is a lack of food or oxygen, stress hormone concentrations in the baby's blood may rise, stimulating the production of estrogen and the beginning of labor. Maternal stress will also prompt the mother's own adrenal glands to produce androgen, just like the fetus. Around the end of a full-term pregnancy, a mother normally contributes about one-third of the androgen that the placenta turns into estrogen. If the mother is under enough stress, her increased production of androgen may cause the level of estrogen to rise far sooner than it should.

One of the first observations of the dangers of stress for pregnancy was by the French physician Adolph Pinard. In 1895 he became concerned by the high rate of prematurity among the young, unmarried women working in the laundry sweatshops of Paris. These female laundry workers had to continue working throughout pregnancy. Their work involved heavy physical labor, lifting wet drapes and other damp laundry, and was psychologically grueling-twelve-hour shifts of hard, repetitive work and few breaks were common at the time. Pinard discovered that keeping these women from working so hard drastically cut the rate of prematurity. His efforts to cut the amount of physical labor during pregnancy were at the forefront of the healthy-baby movement. As we have seen, understanding the adverse effects of excessive physical work led to the idea that exercise in itself, even in moderation, was likely to lead to premature delivery. But as James Clapp and others have shown, it is the stress, rather than the exercise, that is more likely to be responsible for the premature labor.

PREVENTING PREMATURE BIRTH

Since stress is one of the major known causes of prematurity, the number one rule for fighting prematurity is to combat stressors, both in you and the baby. You do this by getting plenty of rest, by getting proper nutrition for the two of you, and by avoiding the physical stresses of infection.

Birth after only thirty to thirty-five weeks of pregnancy is now thought to result primarily from the baby's stress response to insufficient nutrients or oxygen. As I noted in chapter 4, stressed fetuses will pump more steroids out of their adrenal glands—in the same way they do to initiate labor at the end of a full-term pregnancy.

The steroids may be the stressed baby's signal to accelerate the preparations for delivery because things are getting a little difficult inside. Recently, researchers in Australia have suggested that the baby monitors food resources available in the womb, and when food supplies begin to run short, it prompts exactly this sort of "Let's get on with it" signal. Keeping the fetus unstressed and content is important, and can be best accomplished by doing throughout pregnancy all the things that I've recommended:

- Keep eating adequate, nutritionally balanced meals. Morning sickness is usually not a problem late in pregnancy, but the size of the fetus squeezes the stomach and makes it hard to eat much at one time. The solution is to eat small, healthy snacks more often, just as you did in early pregnancy when morning sickness was the problem.

- Keep hydrated. Make sure that you have plenty of fluids. Your blood volume needs to be greatest at the end of the pregnancy to carry the large supply of nutrients. At the same time, you are releasing a greater volume of urine, because you have to expel more waste from a larger baby. This means you have to be especially careful about drinking enough to keep your blood volume up.

- Don't have quintuplets. Twins or triplets are often born a few weeks prematurely because they have to compete for resources with each other and they stretch your uterus to its maximum volume. Stretch destabilizes the muscles and tilts the balance away from the factors that maintain pregnancy and toward the factors that favor labor. Twins occur naturally in 1 out of 80 pregnancies, and triplets occur naturally in 1 out of 6,400 pregnancies. So twins and triplets can be considered a fairly natural part of human pregnancy, and the few weeks of prematurity is entirely within the normal range for the length of pregnancy. Quadruplets, quintuplets, sextuplets, and even septuplets are always born very prematurely for the simple reason that the uterus reaches maximum stretch too early. Competition for space, nutrition, and oxygen between the several fetuses, all jockeying for the best of what is available, likely also accelerates the course of pregnancy.

As I mentioned before, stress in the mother may also set in motion hormonal changes that start labor prematurely. Major life stresses like the death of a close family member, loss of employment, or a less specific stress—such as inadequate emotional support at home or simply a demanding boss—all challenge the maternal stress system. The most damaging sorts of stress are those over which you feel you have no control and which seem to have no resolution in sight. As I noted in chapter 4, decreasing stress requires you to:

- Take positive steps to control your reaction to stressors.
- Get help from those around you to remove stressors.
- Rest often to get rid of physical stress.

Bed rest is one of the most common measures that doctors prescribe when strong contractures become common too early in pregnancy. Physicians also prescribe medications called tocolytics, which interfere with the activity of uterine muscles and stave off the activity of the uterus. One of the problems with these drugs are the side effects.

The most common form of tocolytic drugs mimic some of the activities of adrenaline. As a result, this group of tocolytics often make women feel antsy and jumpy, as if they had taken a lot of caffeine. Strict bed rest combined with tocolytic drugs may be effective in forestalling premature labor in some instances, but this regime can be stressful itself.

An interesting illustration of the importance of social support and stress reduction in preventing prematurity is the story of the tocodynamometer, or toco. The toco is currently the only monitor in wide use for studying the patterns of uterine contraction in women who are threatening to deliver prematurely. It is a simple elastic band that encircles the abdomen and holds a pressure-sensitive disk tight to the mother's abdomen. The uterine wall and its layer of muscles lie right beneath the abdominal wall. When the uterine muscles contract, they bunch up and push the abdominal wall into the pressure-sensitive disk, which sends a signal to a device that can record the size of the contraction. The theory is that when true contractions begin to replace the smaller contractures, labor is in danger of occurring and, if this occurs before the baby is capable of an independent life outside the womb, something should be done to forestall it. Women who are at high risk for premature labor are sometimes hooked up to the machine daily for an hour or two and then the readings are sent to a nurse for analysis. Typically, if the results are good, the nurse will provide praise and encouragement; if there are too many contractions, the nurse will instruct her charge to repeat the measurement, get more bed rest, or check that she is taking the correct amounts of her tocolytic drugs.

What is interesting about this process is the importance of the reassurance provided by the nurse. One study of women at high risk for premature labor divided the women into two groups, one of which had the toco monitoring and daily contact with a nurse. The other group had only the daily contact with a nurse, without using the tocodynamometer. The results show that there is no difference in outcome between the two groups, and it is the contact with the nurse that seems to be beneficial. During the time the pregnant woman is being monitored daily, she feels reassured that a trained professional is watching over her and helping her deal with problems. This is just another example of the importance of social support and the value of feeling more in control of stressors in your life.

Another major cause of premature labor is bacterial infection. Since bacteria are responsible for a large proportion of the premature deliveries for which we can find a cause, the second rule for preventing premature delivery is to detect and protect against infection, particularly bacterial vaginosis. This is true even before pregnancy begins. Some studies suggest that in many cases of premature labor, infection is present at low levels before conception. To fight infection you can:

- Get tested for bacterial vaginosis. If you have had previous episodes of premature delivery, you should get tested every week in the second half of pregnancy. The simplest test for bacterial vaginosis is a pH test. The vagina is naturally somewhat acidic, and the presence of bacteria makes it less acidic. Your doctor can test vaginal pH very simply.

   If bacterial vaginosis is found, it can be treated with antibiotics. Antibiotics have been proven to cut the risk of premature labor in women who have delivered prematurely before. For women who have no history of premature labor, the picture is less clear. Antibiotic treatments can cut the risk of premature labor only slightly in women who have vaginosis but no history of premature labor. Since there seem to be factors (as yet unknown) in addition to infection that are required to initiate labor prematurely, the benefits of fighting vaginosis are not so clear-cut.

- Brush and floss teeth regularly. Recent studies have shown that women with gum disease have perhaps six times the risk of premature labor compared to women with normal gums. These studies have come out at about the same time as other studies showing that gum disease can lead to increased risk of heart disease. What this research shows is that infections in one part of the body can have a significant effect on the health of other parts. Bacteria can either migrate out of the gums to other tissue or set off immunological changes that resonate throughout the body. You should floss regularly as well as brush your teeth. Schedule a visit to your dentist at the beginning of your pregnancy, and then see her again at about the sixth month, when the risk of premature delivery begins to climb.

- Once again, don't smoke. Smoking increases the risk of premature labor by 25 percent. Smoking promotes gum disease because it lowers the normal resistance to bacteria in the mouth. In addition, smokers often have poorer nutrition, which can further lower the body's resistance to infection. As I noted in chapter 6, smoking decreases blood flow to the placenta, which some researchers feel also can set the stage for premature birth.

- It's always a good idea to treat all cuts and blisters promptly, but it is doubly important now that you are pregnant. Don't let a small cut or blister become swollen and infected. Wash all cuts promptly with water and apply antibiotic lotions before covering them with a Band-Aid. A small bit of redness around a cut is extremely unlikely to contribute to premature labor, but if an infection gets started and spreads to surrounding tissue or throughout the body, it can set off the immune and hormonal changes that stimulate contractions.

PREDICTING PREMATURE LABOR

Once labor actually gets going, it is almost impossible to stop. This is good in most cases, since your baby doesn't want there to be any chance that normal labor might stop in the middle of delivery. This quality of labor makes dealing with premature labor harder, however, because even if we invent a drug that dampens down contractions and holds off labor, we will have to be able to administer it very early in the onset of true labor. Heading off premature labor, therefore, is likely to remain difficult until we can both predict which women are likely to start labor prematurely, and treat them well before they start labor in earnest. There have been some recent successes in this regard. Dr. Gillian Lachelin at University College Hospital in London has shown that during the final days of pregnancy, the amount of estrogen in a mother's saliva rises quickly when she is about to go into labor. This intriguing finding has led to one of the most promising biochemical tests for the risk of premature labor. The pregnant woman just thinks of her favorite food, begins to salivate, and collects the saliva in a small tube. The sample is very stable and can be taken to the obstetrician later the same day. A quick laboratory test is available to measure the amount of estrogen in the saliva, and the obstetrician can have the result within a few hours. The best results come when several samples are taken on successive days and a trend can be established.

Another promising test is to measure the molecule fibronectin in secretions taken from the pregnant woman's cervix. Fibronectin is part of the biological "glue" that sticks the fetal membranes to the wall of the uterus. In the first weeks of pregnancy, the membranes around the fetus gradually adhere to the wall of the uterus. At normal delivery, the membranes must peel away from this attachment. When the peeling begins, fibronectin shows up in the cervix and vagina. When the membranes separate from the wall earlier than they should, the demonstration that early separation has begun by showing the presence of fibronectin in the cervical secretions can be a sign that labor is beginning prematurely.

Another indicator of premature labor may be the length of the cervix. As labor nears, the cervix begins to soften and dilate. This dilation makes it flatten and shorten. These changes in the cervix can be measured with an ultrasound. As long as the cervix is longer than 3 centimeters (1.2 inches), labor is extremely unlikely to occur. But if it begins to shorten early, the risk of premature labor increases.

THE FUTURE

The saliva and fibronectin tests have only recently become generally available, and are only the first, most primitive tests of risk from prematurity. In addition, they only show what is happening at one moment of time. What we really need are tools that are similar to those used in the diagnosis of heart disease and the irregular contraction of the heart muscle. In the fifties, sixties, and seventies, there was an enormous effort to develop methods of diagnosis and treatment for heart disease. As a result of years of study, cardiologists have a wide array of tools and tests to help them monitor heart health. By contrast, the obstetrician has only a few meager tools to measure premature labor.

We have to find out if the tocodynamometer and other tests truly can work all the time, or determine under what condition they are not likely to work. The same goes for the testing of drugs that might be able to prevent prematurity. One drug that shows a lot of promise inhibits the action of oxytocin. Several research groups studying pregnant sheep, monkeys, and baboons have shown that this drug is very active against some forms of premature uterine contractions that occur several weeks before the normal end of pregnancy. However, we need more research to determine if the drug would be effective in humans.

One day scientists will find a drug that controls prematurity, just as scientists in the past have found ways to keep alive preemies that would never have survived only twenty years ago. The amazing medical success story of premature babies is how well we have learned how to take care of them. No longer is being born 10 or even 15 weeks early a death sentence. Many premature babies now appear to have no problems stemming from their premature birth as they grow into adolescence and adulthood. I'm sure that medical science will find ways to improve survivability even further, and help tiny precious babies survive and thrive. Premature babies can now do quite well, but there is no doubt that an incubator in the highest quality neonatal intensive care unit is only second best to the swaddling confines of a mother's womb.

The good news is that we have learned a lot more than we

knew even five years ago about the causes of premature delivery and the steps that women can take to avoid it. However, new drugs to block premature delivery, when they do become available, will do nothing to address some of the underlying causes of the problem, such as stress, inadequate nutrition, or bacterial infection. We have the tools to address those problems right now. And we owe it to our children to do better, to find ways to prepare for and provide for them before there is ever any hint that they might be forced from the womb too young.

8

Birth and Beyond: Programming Health During the "Fourth Trimester"

For over nine months, you have been holding and nurturing the child within you, creating a great place to grow where messages from one body pass to the other through the placenta, preparing you both for a new life after birth. After your baby is born this communication continues but becomes more complex. It now falls to your conscious brain to make decisions about the best times for feeding, bathing, and resting. And yet, even though your baby is now outside of you, the intense, often subliminal connection between you continues to create the foundation of his physical and psychological growth. Scientists have found that

- Biologically, birth is not so much a beginning or an end as it is just one developmental landmark along the way.

- While still in the womb, your baby began to learn many of the things that we think of as belonging to life outside the womb—breathing, tasting, learning, hearing.

- You are communicating information about night and day while your baby is still in the womb, so that he can start being in sync with a twenty-four-hour cycle very soon after birth.

- Breast-feeding continues the developmental programming that began in the womb.

- Your stress continues to affect the baby's development.

- Your exercise in the months after delivery can continue to help him develop.

Anthropologists note that in a sense humans are all born "preemies" even after nine months in the womb. Most newborn animals can get up and move around within a few minutes of birth. Newborn guinea pigs, for example, can find and eat solid food for themselves one day after they are born. In contrast, human babies require months before they are mobile and can even reach for food. One reason for such slow development is that we are such large-brained creatures. The human brain more than doubles in size after birth, which is a greater degree of growth after birth than is found in any other animal. If the brain grew at this rate before birth, we would never fit through the birth canal.*

So even though your child is outside the womb, he is continuing to develop in many of the same ways he did in utero. In a sense, the months after birth are a sort of fourth trimester, when you and the baby continue to be physically linked. Like the kan-

---

*Growth of the brain after birth is the key to our superior position in the animal world. As my good friend Professor Dick Swaab, the director of the Netherlands Institute for Brain Research, is fond of saying, "You are your brain." No wonder we humans invest so much in brain development, both before and after birth.

garoo mother with a joey in her pouch, you are responsible for supporting your child, keeping him warm, feeding him, and cradling him as he continues his developmental journey.

If you breast-feed, that physical link is magnified. With breast-feeding, your body continues to be the direct source of your baby's nutrition. Your body processes the food you eat, and supplements it to help the baby grow and to protect him against disease. During this fourth trimester, your diet, exercise, stress control, and exposure to environmental toxins continue to program the lifelong health of your child. Researchers have found that

- The baby's environment after birth influences the quality of neural connections in the growing brain.

- The amount and quality of breast milk is directly related to the mother's diet and stress levels.

- Antibodies in breast milk change the baby's immune system.

- Touching, talking to, and playing with your infant can actually alter neural development and improve brain growth.

- When a mother continues to exercise after giving birth, it tends to help her infant's physical and psychological development.

- Growth-promoting factors are present in breast milk.

- The correct rate of growth after birth is as important as the rate of growth before birth in determining the lifetime health of your baby. Babies who grow either too fast, or too slowly after birth are more likely to have health problems later in life.

In order to promote the healthiest postnatal programming of your baby's health, I recommend you:

- Try to breast-feed your infant for at least the first three months.

- Use stress reduction techniques in your daily life. Some stress is inevitable, but excessive stress takes the joy out of the postnatal experience and impairs your baby's physical and psychological growth.

- Continue to exercise, in one or more of the ways I suggest later. Your exercise helps both you and the baby.

- Enjoy your baby! Playing, hugging, touching, and talking with your baby creates the physical and mental foundation upon which she will build future security and happiness.

GROWING THE BRAIN

Even before birth, the fetus's brain is making and breaking the multitude of connections between nerve cells at an extraordinary rate. Contrary to the old idea that a baby is born as a psychological blank slate, he has actually been learning while in the womb. He was constantly squeezed by contractures of the muscle wall of the womb; he felt the push and bump of your motions, your walk. He swallowed and tasted the amniotic fluid that bathed him. He was listening to the sounds of your voice and the sounds of your stomach and heart. Studies show that newborns can already recognize their own mother's voice, can even recognize the kinds of sounds that she makes in her native language. In the womb, he was already learning to distinguish the sounds that make up words, even though he would not be able to speak those words for many months. And although he couldn't see much of anything in the dark of the womb, he was even learning to see. We now know that small amounts of light penetrate to the womb when the naked stomach is in direct sunlight. Even with no light from the outside, the eyes are creating their own signals, which travel back through the brain and create the brain connections that will be needed to make sense of the visual world. This pattern is happening all over the brain, for all sorts of senses and abilities. One compelling theory of the value of REM sleep and dreams is that they are self-generated signals that the brain uses to wire itself up before birth. As these self-generated signals travel around the brain they help to form connections between nerve cells. It is during the prenatal period that the brain gently molds itself into the shape that it will have to the end of life, and this is also when we have more REM sleep than we do at any other period of our lives. Many of the major neural connections, the organization of neural layers that are so important to information processing, and the main nerve tracts between different areas of the brain are all set up during the time in the womb.

After birth, of course, the environmental stimuli your baby's brain receives are so much greater than they were before birth, and so are the opportunities for learning. In addition, the infant is now living in an atmosphere that contains five times the oxygen that was available in the womb. The higher oxygen concentrations allow the nerve cells to work at a much higher rate and develop the full range of their capabilities. The key to our success as a species is that our brains can receive, assimilate, and organize information so well. The dramatic period of fast growth before birth is crucial to the brain's ability to perform these tasks. With all of this new information flooding into the brain after birth, the nerve cells begin the task of developing the mechanisms to sort and make sense of it all by rearranging the connections between nerve cells. Numerous studies have shown that the ability of any nerve cell to make connections with surrounding cells is improved when that cell is active. In short, I recommend an interesting and stimulating environment because youngsters raised in such an environment develop more capable nervous systems.

The way that most infants get this stimulation is through intense interaction with an adult. Both mothers and fathers have a role to play—as do other members of the extended family. Brightly colored mobiles and motorized toys are fine, but ultimately such things fail to hold an infant's attention for long. The forever changing patterns of pinwheels held by Mom, the surprises of a game of peekaboo, not only stimulate the sensory centers of the brain but also provide highly important stimulation to the emotional centers.

The emotional connection between parent and child, often called bonding, teaches the infant's brain how to make sense of emotional cues. The bonding between parent and child teaches the brain to make nerve connections that allow it to feel warmth and comfort from others. Your child's brain is learning how to process emotions, making sense of interactions he will have with other human beings for the rest of his life. Numerous studies have shown that when monkeys are raised without a warm and comforting mother, they grow up to become cold and distant adults. Other studies show that when mothers stare at their infants impassively for a few minutes, the babies first smile and try to elicit a smile back, then become confused and agitated, and finally become visibly upset. So through either good stimulus or bad, the brain is taught how to interact with other human beings. Emotional connections are being programmed in the neurons.

Although those connections can be altered somewhat by later experience, many of the neural decisions that are made during infancy will have a lasting effect. The strong interaction between parent and child, the emotional bond between the two of you, is a foundation of emotional and physical health for the future. To increase the growth of vital brain connections, nothing can really replace holding, talking to, and playing with your baby. The kind of exaggerated baby talk that seems to come naturally to adults— the cooing, singsong, and repetitious questions and comments— have been shown to be exactly the kind of stimulation he needs to best learn about the world around him.

STRESS

New parents are often unprepared for how highly stressful it is to have a newborn in the house. After all, women find the third trimester of pregnancy difficult enough and after nine months look forward to leaving the physical and emotional issues of pregnancy behind them. But once the baby is born, new mothers may really find themselves overstretched. Feeding, bathing, carrying, and changing the baby, often at irregular hours, day and night, makes the insomnia associated with the third trimester seem like comparatively good sleep. Your newborn may be cooing happily one minute and screaming the next, with no apparent cause for either behavior. Parental life becomes a series of desperate attempts to bring things back to some sort of stable equilibrium. Conversation between spouses often runs along these lines: "Is she hungry? Maybe she's hungry. Perhaps she needs a diaper change. Try rocking her." Even a calm baby may not make you feel any more in control of the situation. Parental confusion reminds me of a story told about actor Laurence Olivier, who brought down the house one night with an especially electrifying performance. A friend went backstage after the show and found the actor distraught. "Why are you upset?" asked the friend. "Your performance was fantastic." Olivier replied: "Yes—but I don't know why." The only good news is that you do learn more as time goes by, and eventually you will forget the bad parts and remember the joyous ones, of which there are plenty.

Meanwhile, a topsy-turvy schedule, sleep deprivation, sensory assault, and an inability to take care of even your most basic needs (oh, for the days when you could just go to the bathroom without making a plan to do so) can easily lead you to feel overwhelmed by stress. Not only is this overwhelmed feeling bad because it lowers your quality of life, it is also dangerous because unaddressed stress can

- Interfere with the parent-child bond, which we have seen is so important to healthy brain development.

- Lead to a feeling of resentment toward your baby.

- Contribute to the development of postpartum depression.

- Reinforce any tendency the baby has developed in the womb toward overreacting to stressful situations.

The stress control techniques that you learned during pregnancy are therefore just as important now as they were before your baby was born. As I mentioned when we reviewed the importance of a thoughtful approach to stress during pregnancy, one of the fundamental stress-fighting techniques is to take control of those things in your life that you can control. You may not be able to control when the baby wakes or is asleep, but you can decide to let the dishes stay dirty and nap when the baby naps. You may not be able to stop your baby from crying in a public place, but you can put her in her car seat and drive, which can divert her attention and stop the crying while also ending your embarrassment by getting her out of earshot. Life with a child is different than it was before, and you learn to do things differently in order to survive psychologically.

Remember also that social support is key for both you and your spouse. Many of the same studies showing that postpartum depression has a negative effect on infants also demonstrate that the women most at risk for this sort of depression have few social supports. As the months go by, the baby will become more manageable, but don't expect to handle everything by yourselves. Couples should get support from other family members or friends. Again, this is a difficult but joyous time, and you need to have help with the difficulties so that you can experience the joy.

Two of the most important supports that family members or friends can provide are these: a little time for yourself, and little time as a couple. You have been physically linked with your baby for over nine months, and once your baby is born you might still feel compelled to spend every moment with her. Don't make that mistake. Yes, you are a mother, but you need a little time apart from the baby (every day, if possible) to repossess your sense that you are a separate person. If you take the time to do things for yourself, you will be a happier person and a better mother for the many other hours of the day that you are with your child.

Besides, your partner also needs some intensive time with his baby. Those responsibilities are also part of the bonding process and will help him feel closer to his child. While he looks after the baby, your time alone may be a good chance to engage in progressive relaxation, meditation, or one of the other stress-reducing exercises you practiced during pregnancy. One of the great stress fighters is, of course, exercise itself. As we shall see, exercise has many other benefits.

Remember also to nurture that crucial bond between you and your partner. Many new parents too easily find themselves falling into the role of job sharing on the baby "project" rather than building the intimate link they have as a couple. Give yourselves time, even if short, to remind yourselves why you wanted to have a baby with your wonderful partner in the first place.

EXERCISE

One of the earliest concerns many women have after giving birth is whether they can do any harm by taking on vigorous physical activity. James Clapp, the Cleveland-based scientist who has done so much research on exercise in pregnant women, recalls that years ago he became concerned when he noticed his patients resuming exercise a few days after labor. He had always been taught in medical school that women should avoid physical stress (climbing stairs, carrying heavy loads) for two weeks after giving birth. The American College of Obstetricians and Gynecologists even now reinforces this sense of caution by recommending that women avoid physical stress for six weeks postpartum. Clapp and other researchers started following the progress of women who resumed exercising within two weeks of giving birth, and found that there did not seem to be a real reason for worry. As long as the activity did not cause pain and did not cause heavy bleeding, there was no harm. This even seemed to be true for the women who had cesarean deliveries. The idea that women should not exercise for a significant time after birth was apparently based on the same lack of data that led to a complete ban on exercise during pregnancy. That advice basically boils down to this: since we don't know if it's bad or not, and we can imagine ways in which it might possibly interfere with recovery from childbirth, women should take the safe course and avoid it. While there is much to commend a "safety first" attitude, it is also important to consider the negative side of missing out on the benefits of moderate exercise.

As with exercise during pregnancy, there are certain things to keep in mind when exercising after pregnancy:

- Don't do any exercise if it causes pain or excessive bleeding. A little spotting is generally considered all right, but make sure you check with your physician about acceptable limits to bleeding.

- Make sure you have adequate breast support, particularly if you are breast-feeding. As with exercise in the last trimester, this may mean wearing two bras or a strong jog bra, as well as supportive clothing while exercising. Don't try to exercise when your breasts are engorged with milk. The best times to exercise might be soon after breast-feeding, when the baby tends to fall asleep and someone else can watch over her while you slip out.

- Stay well hydrated. Getting enough fluids is more important than ever. If you are breast-feeding, you are losing fluid through breast milk and have to drink more water than ever.

- Don't exercise just as a way to lose weight. Getting back to your prepregnancy shape will take time—anywhere from six months to a year. Exercising will help the weight come off rather than stay on, but weight loss after pregnancy will still happen slowly. The goal should be to get out and enjoy the activity, not to improve your performance or silhouette.

- Don't exercise if you have an active infection or abscess in the breast or vaginal area.

- If you have had a cesarean birth or a traumatic vaginal delivery (lots of tearing in the perineal area, for instance) you should be especially careful about beginning to exercise too soon. The recommendation is usually that women with cesarean section delivery should wait six weeks before strenuous activity. It is possible to start exercising sooner, but you should keep close contact with your doctor on this question.

The critical factor that often stops new mothers from exercising, but shouldn't, is time. Getting away to exercise might seem self-indulgent when there is so much to do just to take care of the baby. Don't let yourself fall into that trap. The fact is that parenthood is so demanding that you could use all twenty-four hours in every day and still not get done everything you have to do. You are going to cut corners anyway. I have shown you why exercise (and stress reduction practices) are positive for your child's growth and health. Cut a few more minutes out of your day and do yourself and your baby a lot of good.

NUTRITION

There is no stronger example of the continued connection between mother and child during the fourth trimester than breast-feeding. To allow for rapid brain growth, the baby has now come out of the womb and is breathing on his own. But he is still able to draw strength and sustenance directly from his mother, only now he receives that sustenance through the breast rather than the placenta.

Mother's milk is created in special ducts in breast tissue partly through the action of a hormone called prolactin. Milk is released, or "let down," when the action of a few key nerves in the brain cause the release of another hormone, oxytocin. Oxytocin stimulates the muscle cells that squeeze the milk out of the breasts when your baby sucks. Because it also stimulates the muscle in the womb, oxytocin is often used to initiate or strengthen labor when pregnancy is prolonged or if labor is progressing too slowly. Oxytocin is also released when you have warm and tender feelings toward children, husbands, kittens, etc. It is sometimes called the hormone of love, because it is involved in human orgasm, and because oxytocin can inspire maternal behavior in animals that do not have their own young to care for. The fact that loving feelings prompt oxytocin release explains why an emotional moment can cause breast milk to gush. Emotional stress, on the other hand, blocks oxytocin release and milk letdown. Successful breast-feeding therefore requires a relaxed mother, which is another good argument for exercise and other stress-fighting practices during these months.

There are a multitude of reasons why breast milk is important for a growing child. To begin with, nature has created in breast milk a complete nutritional formula with everything a child needs to grow in a vigorous and healthy way. Carbohydrates and fats provide the energy needed to grow. Fat is particularly important for brain development: during this time the nerve cells receive their covering with a fatty coating that makes signal transmission more efficient and faster. Without these fats, the fatty sheath cannot form correctly. Breast milk has a complete spectrum of proteins to ensure that the body has all the necessary building blocks for creating new tissue. In addition, breast milk contains all the essential vitamins and minerals your baby needs.

Commercial baby formulas can make the same claims, of course, but the many natural benefits of breast milk are impossible to replicate in a commercial product. In breast milk a mother is communicating her immunological memory to her child. Commercial formula, for instance, doesn't contain antibodies, the microscopic proteins that can recognize and help destroy disease-causing bacteria and viruses. The antibodies in breast milk are an archive of the diseases that the mother has ever been exposed to and successfully fought off. Antibodies are especially concentrated in colostrum, the thick, rich substance that the baby sucks out of your breasts in the first days after birth. These antibodies provide a crucial survival edge to newborns, who have been living in a sterile womb and have not yet been exposed to germs. They give him a little assistance in fighting off pathogens until he is able to get his own immune system up and running. The antibodies in your blood and colostrum are the antibodies to the bacteria and other infectious agents in your environment—which is now your baby's environment. So he is getting exactly the right antibodies for his new living conditions.

The antibodies in colostrum and breast milk also help protect the newborn's digestive tract as he gets it working. When our digestive tracts are working well, they contain a huge number of beneficial bacteria that help us digest what we eat. Since the infant's digestive tract starts out sterile, it must be colonized by the beneficial bacteria while excluding the bad bacteria that can cause diarrhea and leave the door open for other pathogens. The antibodies in breast milk help keep out the unwanted bacteria while the beneficial bacteria settle in.

THE BREAST-FEEDING DIET

The quality of breast milk can be affected by what you eat, and even by your stress levels. After all, the milk that you create for your child is only made up of what you eat. It is a product of your body, and if your body is not functioning smoothly, then your breast milk will not be its best. So just as before your baby was born, you still have to eat a well-balanced diet, in order to provide the raw materials for high-quality breast milk.

The breast-feeding diet is similar to the pregnancy diet, except that you have to consume a little more of a few things. One part of a well-balanced diet means consuming enough calories, but not too many. While you ate about 300 more calories per day during the second and third trimesters of pregnancy, the baby is even bigger and growing faster now, so you are going to need to eat about 200 calories more (500 calories per day more than your prepregnancy diet). Post pregnancy is not a time to start eating whatever you feel like again. You've kept in check your desire to overeat during pregnancy, so keep up the good work. The fourth trimester is also not a time to cut caloric consumption way back in order to diet. Don't rush to try taking off the weight you gained in pregnancy. It will happen naturally. Breast-feeding actually helps that process along. During pregnancy, the particular mix of hormones circulating in your blood help you store fat so that you will have energy reserves to pass on to your baby in your milk after birth. So don't worry about the fat, your body will mobilize this source of energy in time. But let your body come back naturally over months under the influence of a healthy diet and moderate exercise.

You will also need to eat more

- Protein. Your baby is now growing faster than ever and you will need to eat a lot of protein. This means about one to two servings more than the two servings per day you might eat before pregnancy.

- Calcium. Growing bones need plenty of calcium, which is why milk is a rich source of calcium. In order for you to put calcium in your breast milk, you are going to need to eat plenty. If you don't, your body will draw calcium out of your bones to feed the baby (your body will lose some calcium after pregnancy anyway). You should drink milk yourself to replace the calcium, as well as eat high-calcium foods.

- Iron. You have to replace any blood you lost during labor, as well as make sure the baby gets iron for his own growing blood supply. Sufficient iron ensures that all the cells in the baby's body get enough oxygen to keep growing.

- Fats. A little fat is a good thing, since milk contains a lot of fat to provide the growing baby with energy. The essential fats that I mentioned in chapter 3 are also important, so make sure that you are getting some fish (which can even be a tuna fish sandwich occasionally) in your diet.

During the fourth trimester, you also have to be especially careful about harmful products you may ingest. Alcohol and drugs pass through your breast and into the baby. If you feel drunk or high, the baby will be getting some of that too. If you are taking painkillers, they will get into the breast milk too. If you have had a cesarean delivery, your doctor will prescribe the correct pain medication. Avoid over-the-counter medications, just as you did during pregnancy. If you drink a lot of caffeine, then your baby will also get the caffeine rush. All the toxins that smoking introduces into your blood will likewise pass into her (not to mention the damage to developing lungs that secondhand smoke causes).

I recommend that you breast-feed for at least the three months that comprise the fourth trimester. Breast-feeding provides real health benefits for your baby and is a natural extension of the emotional and physical link the two of you have had for so many months already.

THE FOURTH TRIMESTER AND BEYOND

The fourth trimester is a time of transition, when you begin to get back your old self, and the baby begins to grow into her new self. At the same time, both of you are still tied together emotionally and physically. Just as you were when you were pregnant, you are helping to build your baby's body, helping her make all the right cells at the right times. Just as importantly, you are helping with her mental health, helping her hardwire the unconscious feelings and memories that will allow her to feel that the world is a secure and loving place.

As your baby grows older, she starts eating solid food, starts forming more permanent memories, starts making the nerve connections that allow her to move on her own from one place to another. The kind of cell growth that leads to permanent, lifelong programming is slowly replaced by a more malleable form of programming. As she grows older, your child learns how to communicate and interact with others, how to be part of social groups, how to choose healthy foods. As she grows older yet, she begins to make time for healthy exercise, in order to replace the boisterous play that was naturally part of her life during childhood. Adulthood will bring the normal stresses of job, family, and daily life. Like all of us, she increasingly will find herself needing to plan how she will stay healthy. She will start getting to the doctor regularly, start counting calories, calculating her daily intake of vitamins and minerals, measuring her heartbeat after exercise. Like all of us, she will need to work at living a healthy life and staying on an even emotional keel.

And yet all of her efforts to create a healthy mind and body are built on the foundation that you help her lay down right now. The health decisions that she makes in the future are shaped by the choices that you made in the first year of her life, during those intimate twelve months she spent growing quickly inside you and then cradled in your arms.

9

The Health of Our World

I know I can always get a rise out of people by telling them that
Title IX, a US law that makes it mandatory to spend as much
money on girls' sports as on boys' sports at schools, has got it all
wrong. I know that until I explain myself, my listeners will think
that I am some sort of throwback Victorian who wants women
only to sit in the parlor with their knitting. Not so. I am delighted
that my daughter, in her thirties, dances several times a week. I
used to run around her high school cross-country course with
her—as I did with my son. No, the point I try to make is that
equal opportunity for active sports for young girls as well as boys
is not just one of providing dollar equality, although that is both
important and fair. The equality aspect is obviously correct, and
shouldn't even be an issue. What's really important about Title IX
is that more sporting facilities for young girls is a health issue.
Young girls and young boys will both determine their likelihood
of fracturing their hips in old age as a result of just how much
strenuous weight-bearing activity they have undertaken during
the first eighteen or so years of their lives. Exercise is vital to
proper maintenance of the body. And most important for the

future health of our society, strenuous exercise is critical to development of the strong heart and good blood vessels that will help a woman have a healthy pregnancy. Because when it comes to the health of the human race, women's health is more important than men's health. Women pass the gift of health from mother to daughter to daughter, down the generations. Men also depend on a good environment during life in the womb, but since men do not have to recreate that environment when it comes time for them to have children, the consequences of maternal health on a good prenatal environment are much more powerful.

The importance of sports programs for girls are only one of the many ways that the repercussions of prenatal programming and prenatal parenting reach far beyond our own families. We all want the best for our children. Most of this book has been about how prenatal programming works and how you can improve your child's lifelong health by starting to parent your child while she is still in the womb. But part of wanting the best for our own children is wanting them to live in the prosperous, healthy, good world. Part of what we need to do to move toward that goal is to help not only our children, but others' children in our town, state, nation, and around the world. For children all over, the concepts of prenatal parenting and prenatal programming have immense implications. They hold incredible power to change our society.

Ignoring prenatal programming holds more ominous implications because, as I like to say, "disease fights forward." What I mean by that is that the greatest effects of disease are in the future. Disease that strikes us in the present generally weakens tissue and makes us more susceptible to further disease and deterioration in later years. A childhood disease can weaken the heart in such a way that the child grows to adulthood and falls victim to congestive heart failure fifty years later.

Or the repercussion may be felt even sooner. In recent years parents have been disturbed to notice that girls are reaching puberty sooner than ever before. It used to be very unusual for ten-year-old, fourth-grade girls to be developing breasts and starting to have periods, but parents have noticed that this is now common in their children or their children's friends. The talk of the PTA meetings has become the focus of scientific research: scientists have confirmed that the average age at which girls start developing breasts in the United States is now ten years old. Further research has confirmed that girls who have a low birth weight—in other words, are subjected to a challenging environment before birth—start menstruating about a year and a half earlier than girls born in the normal weight range. These girls' bodies may simply have learned in the womb to expect a dangerous world out there, one that may not allow them much time to reproduce. These low-birth-weight girls also tend to be shorter once they grow to their full height. Again, the body is concentrating on developing in ways that will help it pass on genes, and is giving short shrift to those attributes that might contribute to a long life.

The fact is that small changes in how we support women's health and pregnant mothers can have a huge effect on every stage of life, from infancy through childhood and puberty to adulthood, middle age, and into old age. When kids die in infancy, when girls can become pregnant in third grade, when a middle-aged husband is hobbled by a weak heart, it puts strains on our society, and we suffer physically, emotionally, and economically.

Right now, in the United States, the major political parties are planning on how to spend a huge budget surplus, and both have promised to lower the national debt. Yet I think that the monetary debt is not the only kind we have to worry about. Difficult conditions in the womb now and in the past have effectively saddled us with a health debt. By not spending relatively small amounts of money to help parents provide a better home for children in the womb, we are only promising to pay large amounts for health care decades from now. We already pay $600 billion per year for health care in the United States. What will that figure be fifty years from now, when the population is much greater and scientists have dreamed up all sorts of wonderful, expensive miracle treatments? We can pay now or pay later. And more important than the money we can save are the lives that we can save, the suffering that we can eliminate.

The story of prenatal programming started nearly a century ago, in hospitals in Sheffield, England; in Mysore, India; and at the London Missionary Society and the Rockefeller Hospital in Beijing. Nurses carefully collected and archived their detailed records on mothers and their babies long before the advent of megahertz computers and powerful databases to store, sort, and tabulate. They wrote down the details of the new arrivals in leather-bound journals or on plain sheets of paper, unable to conceive that their routine but painstaking attention to detail would enable future generations to begin to see how they can pass good health down the chain of life. Now scientists have shown us how important those first months of life really are, and it's time to apply those lessons. It is when we apply knowledge that we can create positive changes in our lives.

Every time I see huge earth-moving equipment on the freeway, I am reminded of the old ways of producing engineering works. I see mountains of earth with antlike humans scurrying up and down with small baskets of dirt. In some parts of the world, human power is still the main way of getting large projects completed. However, it is clear that to accelerate productivity and provide greater efficiency, we need brains rather than brawn. We need to apply our knowledge and follow intelligent courses of action in order to provide an energized, productive society. This is just as true, or more so, in medicine. In the 1930s we had a very effective therapy for treating late-stage polio: the iron lung. When the muscles became so paralyzed that the patient could no longer breathe, he was placed inside an iron tank that would, like a huge vacuum cleaner, pull air into the lungs by sucking air out of the surrounding tank. But we didn't solve the problem of polio by building a better iron lung. We cured polio by understanding that polio is transmitted by a specific virus, and learning how to immunize against polio by administering a weakened strain of that virus.

I hope that as the story of prenatal programming goes forward into the future, we will start to apply what we know to all families. We will be able to take what is the science of prenatal programming and turn it into the practice of prenatal parenting. But the first stage is for people to hear the story and understand the message.

There are many signs that this is happening, but I find particular hope in an unusual invitation I recently received. The invitation was to speak to a gathering about prenatal programming, but that is not unusual in itself. What was different about this invitation was that it asked me to speak to the World Affairs Council; this organization regularly invites people to speak before a large audience of council members on issues affecting international relations and policy. Speakers tend to be ambassadors to the United Nations or policy wonks from the Agency for International Development. The week before my appearance before the council, the speaker had been Richard Butler, the man in charge of the UN monitoring of Iraq's disarmament after the Persian

Gulf war. I am happy to say that the audience seemed interested and seemed to understand the public policy implications of prenatal programming for building stronger societies and stronger economies, one family at a time. At least, that is my hope and my dream.

Index